Controlling Silver Discharges in Wastewater

Prepared under the direction of the **Pollution Prevention Committee**
Margaret Nellor, *Chair*
Karl Sittner, *Project Chair*

1994

Water Environment Federation
601 Wythe Street
Alexandria, VA 22314–1994 USA

Library of Congress Cataloguing-in-Publication Data

Technical guidance manual for silver discharges/WEF Pollution Prevention Committee.

p. cm.

"Draft."

"March 1994."

Includes bibliographical references and index.

ISBN 1-881369-97-8

1. Sewage—Purification—Silver removal. 2. Factory and trade waste—Management. I. WEF Pollution Prevention Committee. TD758.5.S44T43 1994

628.1′683—dc20 94-300 03 CIP

Library of Congress Catalog No.

ISBN

Printed in the USA 1994

Special Publications of the Water Environment Federation

The WEF Technical Practice Committee (formerly the Committee on Sewage and Industrial Wastes Practice of the Federation of Sewage and Industrial Wastes Associations) was created by the Federation Board of Control on October 11, 1941. The primary function of the Committee is to originate and produce, through appropriate subcommittees, special publications dealing with technical aspects of the broad interests of the Federation. These manuals are intended to provide background information through a review of technical practices and detailed procedures that research and experience have shown to be functional and practical.

IMPORTANT NOTICE

The contents of this publication are for general information only and are not intended to be a standard of the Water Environment Federation (WEF).

No reference made in this publication to any specific method, product, process, or service constitutes or implies an endorsement, recommendation, or warranty thereof by the Federation.

The Federation makes no representation or warranty of any kind, whether expressed or implied, concerning the accuracy, product, or process discussed in this publication and assumes no liability.

Anyone using this information assumes all liability arising from such use, including but not limited to infringement of any patent or patents.

PREFACE

Controlling Silver Discharges in Wastewater is primarily designed as a guide in establishing a source control program for silver discharges. The publication identifies the types of industries discharging silver-bearing wastewater. It discusses the nature and quantities of the pollutants discharged by these industries. A discussion of both conventional and innovative control technologies is presented.

The process of planning and organizing a source control plan is discussed in this manual, as well as methods of identifying industries and gaining their involvement, setting policies and goals, and securing management, employee, and policymaker (elected officials) support. It describes procedures for setting up benchmarks and methods of evaluating program success. Examples of successful programs are presented.

Principal authors for this publication are

Robert Cappel
Dirk R. Dunbar
Chris Ferrigan
Adriana Renescu

Additional review was provided by

Joanne DiCaro
J. Kevin Farmer
Kris P. Lindstrom

Authors' and reviewers' efforts were supported by the following organizations:

City of Boca Raton, Florida
City of Palo Alto, California
City of San Bernardino Water Department, California
Eastman-Kodak, Rochester, New York
James M. Montgomery Consulting Engineers, Inc., Walnut Creek, California
K.P. Lindstrom Inc., Pacific Grove, California

Ministry of Environment and Energy, Toronto, Ontario, Canada
O'Brien & Gere Engineers, Inc., Syracuse, New York
Orange County Sanitation District, Fountain Valley, California
Rochester Midland Corp., Rochester, New York

Federation staff project management was provided by J. Robert Schweinfurth. Technical editing was provided by Matthew Hauber.

CONTENTS

Section **Page**

CONTENTS **(continued)**

Section **Page**

CONTENTS (continued)

LIST OF TABLES

LIST OF FIGURES

LIST OF FIGURES (continued)

INTRODUCTION

The National Pretreatment Program has charged wastewater treatment plants (WWTPs) with the responsibility of controlling the discharge of potentially toxic or hazardous substances to the environment. The changing nature of American industry, increasingly restrictive sludge disposal regulations, and tighter discharge permit limits on various water-quality-based toxics place added emphasis on WWTPs to control the discharge of toxic or hazardous pollutants by industrial users. As a result, the role of pretreatment in controlling toxic or hazardous pollutants is continually increasing.

One hazardous substance coming under closer scrutiny is silver. To comply with a mandate to reduce silver discharges, the development and implementation of a source control program for silver may be necessary.

The Need for a Source Control Program

The need for a program to control silver is generally determined by the regulating agency. This determination is driven by three basic factors in the WWTP operation:

- Discharge requirements as defined by the WWTP's discharge permit,
- Special discharge requirements to estuaries and bays as determined by regional water quality boards or the U.S. Environmental Protection Agency (U.S. EPA), and/or
- Silver limits in operations residuals such as sludge or ash.

Once such a need has been determined, planning and organizing a source control program can begin. Elements of a successful program include

- Securing WWTP management and staff commitment,
- Setting policies and program goals,
- Identifying silver-discharging sources and establishing an advisory group with their participation,
- Setting up the program with the participation of the advisory group,
- Implementing the program and monitoring its results, and
- Benchmarking program results to existing successful programs.

A source control program for silver can become an integral part of the WWTP's operation. Once inculcated into the WWTP process, it can serve as an invaluable check on overall efficiency of the plant's operation as well as an aid in meeting legal requirements.

THE PHOTO-PROCESSING INDUSTRY

Photographic films and paper are used to diagnose medical problems and structural defects; document, record, and transfer information; and record personal events. While the manufacture of photographic film and paper occurs at only a few sites around the world, the processing of these materials occurs at numerous facilities. Few (<1%) of the total number of these facilities discharge more than 95 m^3/d (25 000 gpd) of process wastewater. Most photoprocessing facilities, therefore, are not regulated as significant industrial users (SIUs).

Photoprocessing facilities that could be classified as SIUs include most hospitals; a few diagnostic clinics, printers, and photofinishers; and major motion picture film processors. Large photoprocessors, representing about 9% of all photoprocessing facilities, typically use between 38 and 95 m^3/d (10 000 and 25 000 gpd) of process wash water and produce between 380 and 950 L (100 and 250 gal) of silver-rich processing solution per week. Because of silver's value as a precious metal, large photoprocessing facilities have focused substantial efforts on silver-recovery operations for many years. Using a combination of electrolytic and metallic replacement technologies, these facilities have typically recovered more than 99% of the silver from their silver-rich processing solutions.

Small- to medium-size photoprocessors, representing approximately 90% of the total number of photoprocessing facilities, include doctor's offices; neighborhood clinics; schools; portrait studios; minilabs (1-hour service operations in shopping malls and stores); custom print shops; professional processing labs; small microfilm and motion picture processors; home hobbyists; and a number of municipal, state, and federal facilities where some photoprocessing is done.
Medium-size facilities typically use between 4 and 38 m^3/d (1 000 and 10 000 gpd) of process wash water and produce between 38 and 380 L (10 and 100 gal) per week of silver-rich solution. Small facilities typically use less than 4 m^3/d (1 000 gpd) of process wash water and produce less than 38 L (10 gal) per week of silver-rich solution. An example of the number of different types of photoprocessors is shown for Phoenix, Arizona, USA, in Table 1.

THE PHOTOGRAPHIC PROCESS

Silver-based photographic materials consist of solid crystals of silver chloride or silver bromide suspended in gelatin and coated on a film or paper support. The processing of photographic film and paper consists of

Table 1 Number of photoprocessing facilities, Phoenix, Arizona (data supplied by the Center for Government Research, Rochester, N.Y.).

Facility type	Number of facilities by size (in number of employees)			
	Small (1 to 19)	Medium (20 to 49)	Large (50 to 499)	(More than 500)
Health care				
Hospitals[a]	14	1	61	32
Dentists	1 422	16	1	0
Doctors	915[b]	122	47	2
Veterinarians	278	9	2	0
Chiropractors	515	6	0	0
Commercial				
Minilabs	184	0	0	0
Photofinishers	0	5	5	0
Professional labs	129	9	2	0
Motion picture	1	0	0	0
Microfilm	12	3	1	0
Graphic arts	783	101	47	3
Noncommercial				
Schools	0	0	5	7
Police departments	4	3	18	11
Heavy construction	269	57	58	4
Transportation	14	2	44	7
Fabricated products	18	5	42	10
Finance/insurance/real estate	0	0	22	10
Jewelry/silverware/plated ware	47	1	0	0

[a] Does not include university, college, or public hospitals.
[b] Includes offices of podiatrists, osteopaths, and 10% of all medical doctor offices.

three basic steps: developing the image, removing the silver, and stabilizing the image by rinsing residual thiosulfate and silver–thiosulfate complex out of the emulsion layers with water or using a stabilizer solution for washless processing.

During exposure to light, small, invisible specks of metallic silver are formed on the surface of the silver halide crystals. The size of the metallic silver speck is proportional to the amount of light hitting the crystal. These small specks must be amplified, or developed, to view the image. This is accomplished by immersing the film or paper in a water solution containing a mild chemical reducing agent, called the *developing agent.* The developing agent reduces more silver ions where the metallic specks have formed from light exposure: the larger the metallic speck, the more silver metal the developing agent will produce at that site. Thus, areas of high exposure produce greater amounts of metallic silver and high densities, while areas of low exposure produce small amounts of metallic silver and low densities. The degree of development produced is controlled by varying the time of immersion, activity or pH of the developing solution, and concentration of the developing agent. As the developing agent reduces the silver halide, it becomes oxidized and cannot be used further in the amplification process. In color films and papers, however, the oxidizing developing agent is used to form the dyes of the final image. Therefore, the developing agent must be replaced or replenished after the film or paper is processed.

After development, if the film or paper were exposed to light, the silver halide crystals still present in the gelatin coating in areas of little or no exposure would continue to form metallic silver until the surface turned black. Thus, the image must be stabilized or *fixed.* This is done by immersing the film or paper in a water solution of thiosulfate called a *fix.* Thiosulfate forms a strong, soluble complex with silver, dissolving the silver halide crystals.

In black-and-white films and papers, such as those used in X-rays, graphic arts, microfilm, and black-and-white motion picture and professional photography, the image is metallic silver. However, in color films and papers, such as used in amateur, professional, motion picture, and graphic arts photography, the black metallic silver image must be removed, or *bleached,* to view the color dyes. This is accomplished by immersing the film or paper in a water solution of a mild oxidizing agent with a halide, such as chloride or bromide, that converts the metallic silver back to crystals of silver halide. An iron complex of ethylenediaminetetraacetic acid (EDTA) is typically used for this process in a solution called a *bleach*. After bleaching, the film or paper is immersed in the fix to dissolve the silver halide crystals. In some paper processes, the iron–EDTA salt and thiosulfate are combined into the same solution, called a *bleach-fix*.

After the use of the abovementioned solutions, the soluble silver–thiosulfate complex and the other residual chemicals are removed from the gelatin coating by washing with water.

After washing, the film or paper is immersed in a final water solution containing a small amount of wetting agent and occasionally a gelatin hardening agent before it is dried.

PROCESSING SOLUTIONS AND VOLUMES

Developer solutions typically contain less than 1% reducing agent by weight. For color developers, the reducing agent is a substituted para-phenylenediamine. The oxidized para-phenylenediamine formed during development reacts with another chemical to form the image dyes and remains in the gelatin layer. For black-and-white developers, the reducing agent is usually hydroquinone, a sulfonated hydroquinone, or ascorbic acid. The highest developing agent concentrations are found in some high-silver medical X-ray processes, for which the hydroquinone concentration can be as high as 3% by weight. In black-and-white processing, the oxidized developing agent remains in the developer solution.

The other major components of a developer solution consist of a pH buffer (typically carbonate) to maintain pH in the range of 9 to 11, a calcium sequestrant (typically EDTA or a metaphosphate), and an antioxidant (typically sulfite or a hydroxylamine). These compounds are present in concentrations generally less than 1% by weight, in the 1- to 10-g/L range. Developer solutions contain more than 90% water by weight.

Because developing agents oxidize on exposure to air, the developing solutions are unstable and degrade with time. The oxidized ingredients must be replaced to maintain consistent results. As new solution is added to the developer container, an equivalent volume is discharged from the processor. This spent developer solution contains more than 90% water, with a few grams per litre inorganic salts (carbonate and sulfate) and portions of developing and sulfonated developing agents.

Bleach solutions contain 10 to 30% iron–EDTA complex, typically the ammonium salt of ferric EDTA or ferric propylenediaminetetraacetic acid (PDTA). Ammonium salts are used because they are more soluble and transport through the gelatin layers faster than sodium or potassium salts. Thus, lower concentrations can be used. The bleach also contains 5 to 10% acetic acid and an acetate salt to buffer the pH in the 4 to 6 range, 5 to 10% bromide salt, and a small amount (<1%) of sodium or potassium nitrate, which is used to prevent corrosion of the processing tanks. As it oxidizes the metallic silver, the ferric complex is reduced to ferrous salt and must be regenerated or replaced. As this solution is replenished, the overflow will contain 65 to 85% water, ferrous and ferric EDTA or PDTA complexes, and inorganic salts such as bromide, nitrate, and ammonium ion.

Fix solutions contain 65 to 85% water, 10 to 30% ammonium thiosulfate, and 5 to 10% sulfite salt that acts as an antioxidant. Fix solutions are not stable; exposure to air slowly degrades thiosulfate to elemental sulfur. The fix solution also contains the soluble silver–thiosulfate complex, and a seasoned solution should contain between 1 000 and 5 000 mg/L silver. As the fix is replenished, the overflow is collected for silver recovery.

After the fix, the films or papers are immersed in a series of wash tanks to remove the silver–thiosulfate and other residual chemicals from the gelatin layers. This washing imparts stability to the image. In one amateur film and paper process, the water washes are replaced by a replenished stabilizer. This stabilizer contains citrate salts and polyvinylpyrrolidone to complex or react with the residual chemicals and provide image stability. As this solution is replenished, the overflow is collected for silver recovery. While the image stability is not as good as that provided by water washing, it is good enough for most amateur photographers who keep their negatives and prints in boxes and albums.

Stabilizers or final rinse solutions are 99% water except for washless color processes. These solutions contain a wetting agent to prevent water spotting during drying. In some color film stabilizers, a small amount of formaldehyde ($<0.2\%$) is present to harden the gelatin layer or stabilize an image dye.

Developer replenishment rates typically range between 5 and 30 mL/sq ft for paper processes and 15 and 100 mL/sq ft for film processes. Bleach and bleach-fix replenishment rates vary from 5 to 30 mL/sq ft for color processes. Fix replenishment rates range from 15 to 100 mL/sq ft for film and paper processes. Stabilizers are replenished at rates between 10 and 30 mL/sq ft. Wash waters are replenished at 200 to 1 000 mL/sq ft for each wash tank in the process. The average volumes of different-size photoprocessing facility operations are shown in Table 2. Silver-rich solutions include fix, bleach-fix, and washless stabilizers. Low-

Table 2 **Average daily volume of silver-containing fluids produced by different-size photoprocessing facility operations (data supplied by the Silver Coalition, Harrison, N.Y.).**

Facility	Silver-rich solutions, gpd*			Low-silver solutions, gpd		
	Small	Medium	Large	Small	Medium	Large
Health care						
Dental offices	0.1	0.2	0.4	480	1 200	1 920
Hospitals	20	40	80	2 600	5 200	10 400
Medical professionals	0.2	1	5	100	500	1 000
Commercial						
Microfilm	0.1	0.3	50	15	75	3 750
Printers/graphic arts	1	2	20	255	450	4 500
Minilabs: washless (80%)	2	3	5	1	1	2
Minilabs: washwater (20%)	1	0.5	1	45	90	180
Photofinishers/professional labs	10	100	260	1 300	13 000	32 500
Motion picture	25	50	2 000	1 000	2 000	80 000
Government facilities						
Police departments	0.2	0.4	2	25	50	125
Hospitals	20	40	80	2 600	5 200	10 400
Printing offices	1	2	20	225	450	4 500
Microfilm processing	0.1	0.3	50	15	75	3 750
Industrial X-ray	5	10	20	500	1 000	2 000
Professional photography	1	5	10	125	650	1 300
Higher education						
Medical program/hospital	20	40	80	2 600	5 200	10 400
Graphic arts	1	2	20	225	450	4 500
Dentistry	—	—	0.1	—	—	1 920
Veterinary/chiropractic	—	—	5	—	—	1 000
Photography	1	5	10	125	650	1 300
Industrial						
Printed circuit boards	1	2	20	225	450	4 500
Transportation	5	10	20	500	1 000	2 000
Heavy construction	5	10	20	500	1 000	2 000
Oil and gas/forestry	1	5	10	125	650	1 300
Photographic equipment	1	5	10	125	650	1 300
All others	—	5	10	—	500	1 000

* gpd × (3.785×10^{-3}) = m^3/d.

silver solutions include developers, bleaches, stabilizers used after washes, and wash waters. Dental office low-silver volumes also include water from dental chair discharges.

The total process effluent from facilities using wash water in the process has the characteristics listed in Table 3. The total process effluent from facilities using the washless processes has the characteristics listed in Table 4. While the washless concentrations are many times higher than the wash process, the total loading is equivalent.

Table 3 Conventional photoprocessing effluent characteristics (data supplied by the Silver Coalition, Harrison, N.Y.).

Parameter	Concentration range, mg/L
Temperature	80–110°F[a]
pH	6.5–9.5 units
Biochemical oxygen demand (5-day)	200–3 000
Chemical oxygen demand	400–5 000
Total dissolved solids	300–3 000
Total suspended solids	< 5–50
Ammonia nitrogen (NH_3–N)	20–300
Total Kjeldahl nitrogen (TKN)	30–350
Thiosulfate	100–1 000
Sulfates	50–250
Silver[b]	< 0.1–5
Iron	< 10–100
Zinc	< 0.75

[a] (°F−32) × 0.555 6 =°C.
[b] After silver recovery.

SILVER-RECOVERY TECHNOLOGIES

There are three basic technologies available to photoprocessors for recovering silver from spent silver-rich photoprocessing solutions: electrolysis, metallic replacement, and chemical precipitation. Ion exchange can be used to recover silver from spent wash waters. Reverse osmosis and evaporation are concentration technologies, not silver-recovery technologies (see Figure 1). Technologies used in metal finishing such as electrowinning cannot be used for spent photoprocessing solutions or wash waters because of equipment fouling and heat decomposition of some of the process

Table 4 Washless photoprocessing effluent characteristics (data supplied by the Silver Coalition, Harrison, N.Y.).

Parameter	Concentration range, mg/L
Temperature	< 95°F[a]
pH	6.5–9.0 units
Biochemical oxygen demand (5-day)	5 000–15 000
Chemical oxygen demand	10 000–35 000
Total dissolved solids	30 000–90 000
Total suspended solids	10–50
Ammonia nitrogen (NH_3–N)	6 000–10 000
Total Kjeldahl nitrogen (TKN)	8 000–13 000
Thiosulfate	20 000–25 000
Sulfates	3 000–4 000
Silver[b]	< 5–50
Iron	1 400–2 000
Zinc	< 2

[a] (°F–32) × 0.555 6 =°C.
[b] After silver recovery.

chemicals resulting in noxious odors. Some pretreatment programs apply the categorical limits of 0.24 or 0.43 mg/L for direct-discharge metal finishers to photoprocessors who are not categorical and are indirect dischargers. The best available technology limits for photoprocessors who are direct dischargers are listed in 40 CFR 459.

Electrolysis

Electrolytic recovery is an effective environmental option for silver recovery because the equipment can be reused many times and no additional chemicals are released as a result of the recovery operation. An electric current reduces the silverthiosulfate complex and plates almost pure silver metal out on a piece of stainless steel or other material called the *electrode* or *cathode*. There are two basic types of equipment, one in which the cathode rotates in the solution and one in which the solution flows around a stationary cathode. Both types of equipment are capable of recovering over 90% of the

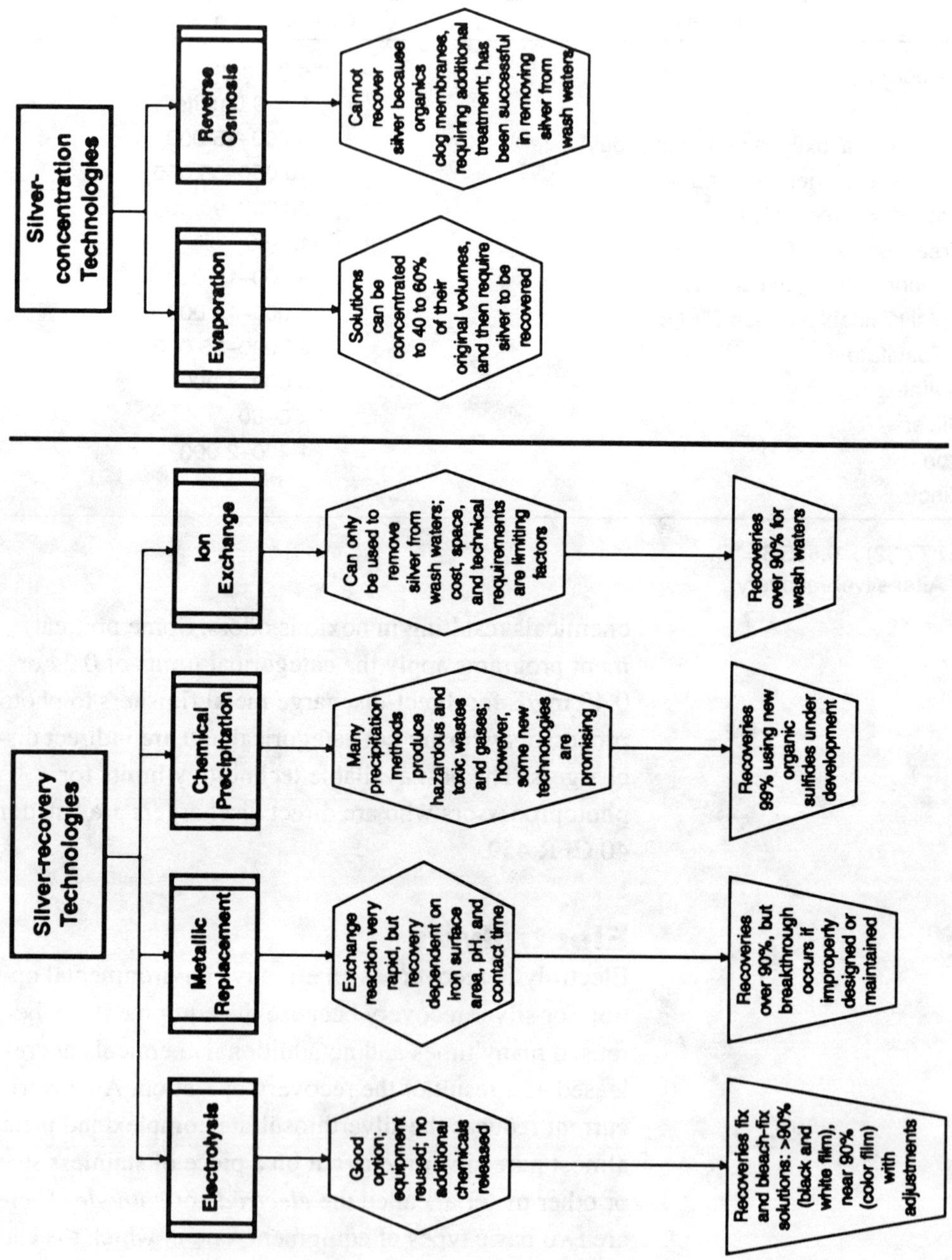

Figure 1 **Silver-recovery and -concentration technologies.**

silver from the spent silver-rich solutions. However, equipment in which the electrode or cathode rotates tends to require more maintenance.

Spent fix solutions from black-and-white processes are easy to desilver electrolytically and require little, if any, adjustment. Recovery efficiencies of over 90% can be achieved. Attempts to extend the electrolysis time or increase the current density typically result in the silver that is plated on the cathode being coated with a black sulfide precipitate, which substantially reduces the efficiency of recovery.

Spent bleach-fix and fix solutions from color processes that contain iron-complex oxidizing agents are not as easy to desilver electrolytically. Higher-current densities and longer times are required, with occasional pH adjustments. Under acidic pH conditions (pH <7.0), the iron–EDTA complex used in bleaches and bleach-fixes by small color photoprocessing facilities oxidizes the silver plated on the electrode and moves it back into solution. As the pH becomes more basic (pH >7.0), the oxidizing power of this iron–EDTA complex is reduced and the plating of silver onto the electrode becomes more efficient. The optimum pH for silver-rich solutions containing iron is pH 8.0 to 8.5. With proper equipment and pH adjustment, silver-recovery efficiencies approaching 90% can be achieved from these solutions using electrolytic recovery.

Metallic Replacement

Elemental iron undergoes an oxidation-reduction reaction with silver–thiosulfate to produce ferrous ion and metallic silver. This reaction can be accomplished by pouring the spent fix or bleach-fix through a container of steel wool, iron particles, or an iron-impregnated resin. The exchange reaction is rapid but dependent on the contact of the silver–thiosulfate complex with the iron surface. Thus, the iron surface area, the contact time, and the solution pH are the major variables influencing recovery efficiencies. Properly designed silver-recovery cartridges, sometimes referred to as chemical-recovery cartridges, can recover more than 90% of the silver from spent silver-rich solutions.

As the recovery cartridge is used, the active surface area is reduced and small channels begin to develop in the iron substrate. The efficiency of recovery gradually diminishes until breakthrough occurs. A small pump or flow restricter to control the flow of the spent silver-rich solutions through the cartridges will improve the recovery efficiency. Two cartridges connected in series, with a valve in between to allow testing, prevents the discharge of silver-rich solutions to the drain. This properly maintained 2-cartridge system should ensure at least 95% silver recovery from the spent silver-rich solutions from small photoprocessors.

Chemical Precipitation

The silver–thiosulfate complex can be reduced by chemicals such as borohydride or amineboranes. However, these chemicals can pose hazardous situations under these circumstances and are not recommended for photoprocessors.

Silver can also be precipitated as a sulfide by adding sodium sulfide, or as an insoluble complex using organic sulfides. Inorganic sulfide use can pose problems with releases of highly toxic hydrogen sulfide gas and requires careful control of reagent photoprocessing facilities. However, some organic sulfide precipitation processes currently under development may provide at least 99% recovery efficiencies.

The newest precipitation technology uses a sodium salt of trimercapto-*s*-triazine (TMT). This technology can be used as a primary recovery system in small photoprocessing facilities or secondary recovery system following electrolytic recovery for larger facilities. The insoluble silver–TMT complex is removed from solution by settling and filtration.

Ion Exchange

Two types of ion exchange technology exist. Conventional ion exchange relies on attracting the silver–thiosulfate negatively charged complex to positively charged sites on the ion exchange resin. When the positively charged sites are filled, breakthrough occurs and the resin is regenerated by removing the silver–thiosulfate complex with a concentrated negatively charged ion solution.

In situ ion exchange relies on converting the silver–thiosulfate to a silver sulfide precipitate in or on the resin beads. These resin beads are typically sent to a silver refiner or reclaimer, where they are incinerated to recover the silver.

Ion exchange cannot be used for recovering silver directly from spent fix and bleach-fix solutions. These solutions can strip silver from either resin and actually result in more silver being discharged from the resin column than present in the feed solution. The acidification process required for *in situ* precipitation can also release sulfur dioxide from these solutions, presenting a potential hazard. This acidic solution must also be neutralized before disposal to the sewer.

Thus, ion exchange lends itself only to the recovery of silver from wash water. While ion exchange can remove more than 90% of the silver from wash water, its cost, space, and technical requirements preclude its use in most photoprocessing operations. Additionally, the environmental impact of this technology is negative when compared to the small amounts of silver it recovers. However, it could be a valuable technology in areas where wash water reuse and conservation are especially important.

Reverse Osmosis

Reverse osmosis is a concentration technology that requires additional treatment to recover silver. Reverse osmosis has been unsuccessful in removing silver from photoprocessing silver-rich solutions because the organics present in these solutions tend to clog the membranes quickly. Reverse osmosis has been used successfully to remove silver from some process wash waters. However, the cost, space, and energy requirements of this technology make it unacceptable for application in most photoprocessing facilities.

Evaporation

Evaporation represents another way to concentrate silver-rich processing solutions. Silver-rich processing solutions can be concentrated to 40 to 60% of their original volume. This technology requires additional treatment to recover the silver.

PLANNING AND ORGANIZING A SOURCE CONTROL PROGRAM

The development and implementation of a source control program for the photoprocessing industry by a WWTP has several phases and begins with the decision that such a program is necessary. The concluding phase of the program consists of implementation and provisions for the maintenance of the program. The elements of developing or constructing a program include identifying dischargers, developing strategies, coordinating and collaborating with the stakeholders, and marketing the program. The implementation of the program proceeds along established milestones and progress is measured by benchmarks and the attainment of objectives. The program should include a maintenance system that ensures goals and objectives of the program are maintained after implementation.

The source control program for photoprocessing targets industries that generate silver in the processing and development of photographic film. Although only one compound, silver, is targeted, because of the variety of dischargers and their sizes, the program should be tiered according to groups and types of dischargers. In addition, surveys performed to evaluate source control programs for photoprocessors show that the bulk of dischargers fall in the small, commercial category (AMSA, 1993). Figures 2 and 3 present flow charts of the steps and decision points in the development of a source control program for photoprocessors and silver control. Each of these steps is discussed below.

Establishing the Need and Setting the Goals and Objectives

The basis of the source control program is a determination by the regulating WWTP that silver discharges to its facilities must be reduced, and that a source control program would be appropriate and effective in attaining the proposed reduction. Only after the need and the basis for the need have been established can the planning and development of the source control program for silver take place.

Generally, the need for a silver discharge control program is driven by three basic requirements in the WWTP operation: discharge limits defined by the WWTP's National Pollutant Discharge Elimination System (NPDES) discharge permit; special discharge requirements to estuaries and bays mandated by regional water quality control boards or U.S. EPA (for example, San Francisco Bay area WWTPs are subject to special discharge limits to the bay because of shellfish contamination); and silver limits in operations residuals such as sludge. These elements form the basis and the legal framework for the development and implementation of the regulatory program to control the discharges of silver-bearing wastewater to the sewer.

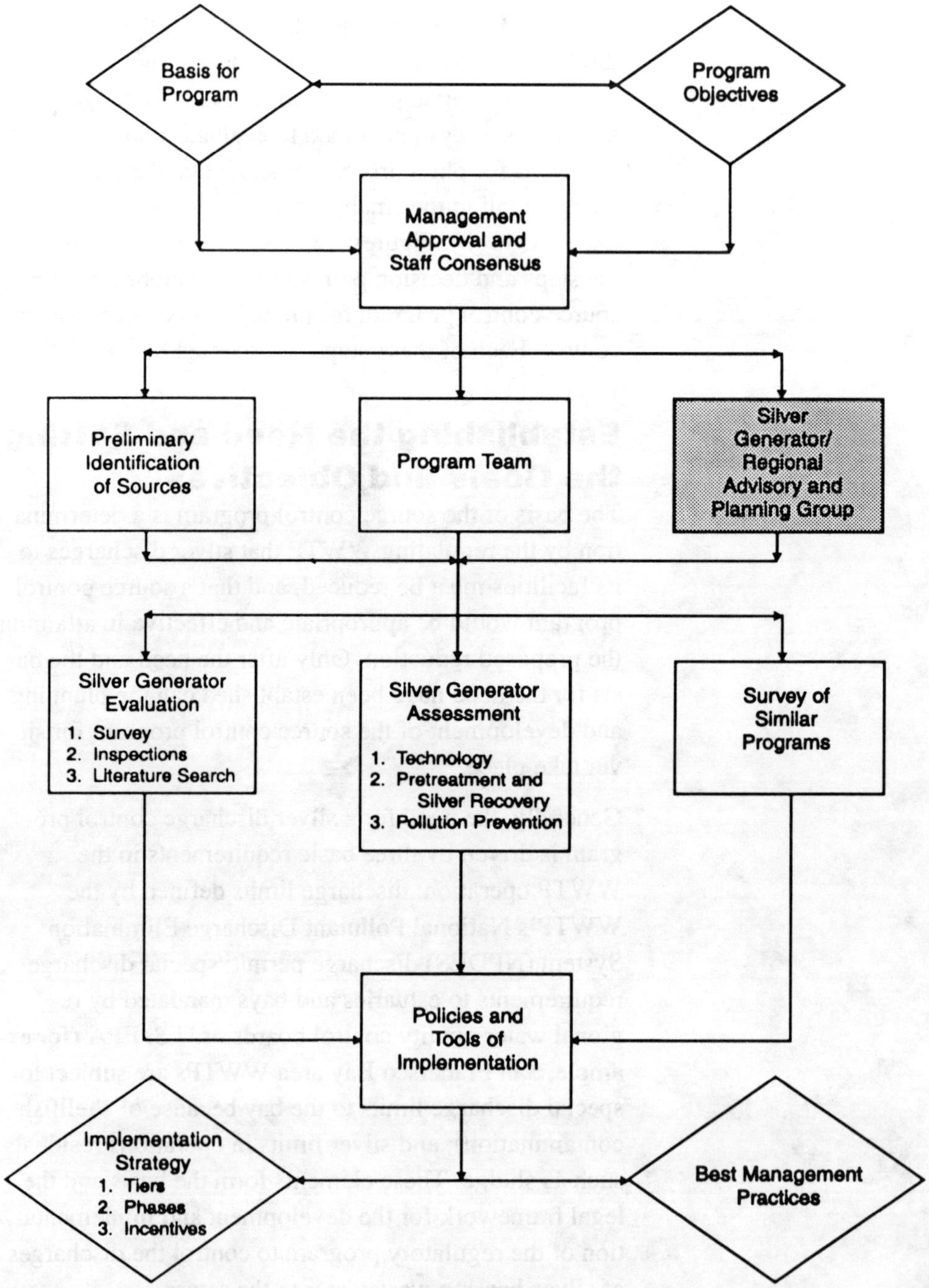

Figure 2 Silver source control program development, phase I.

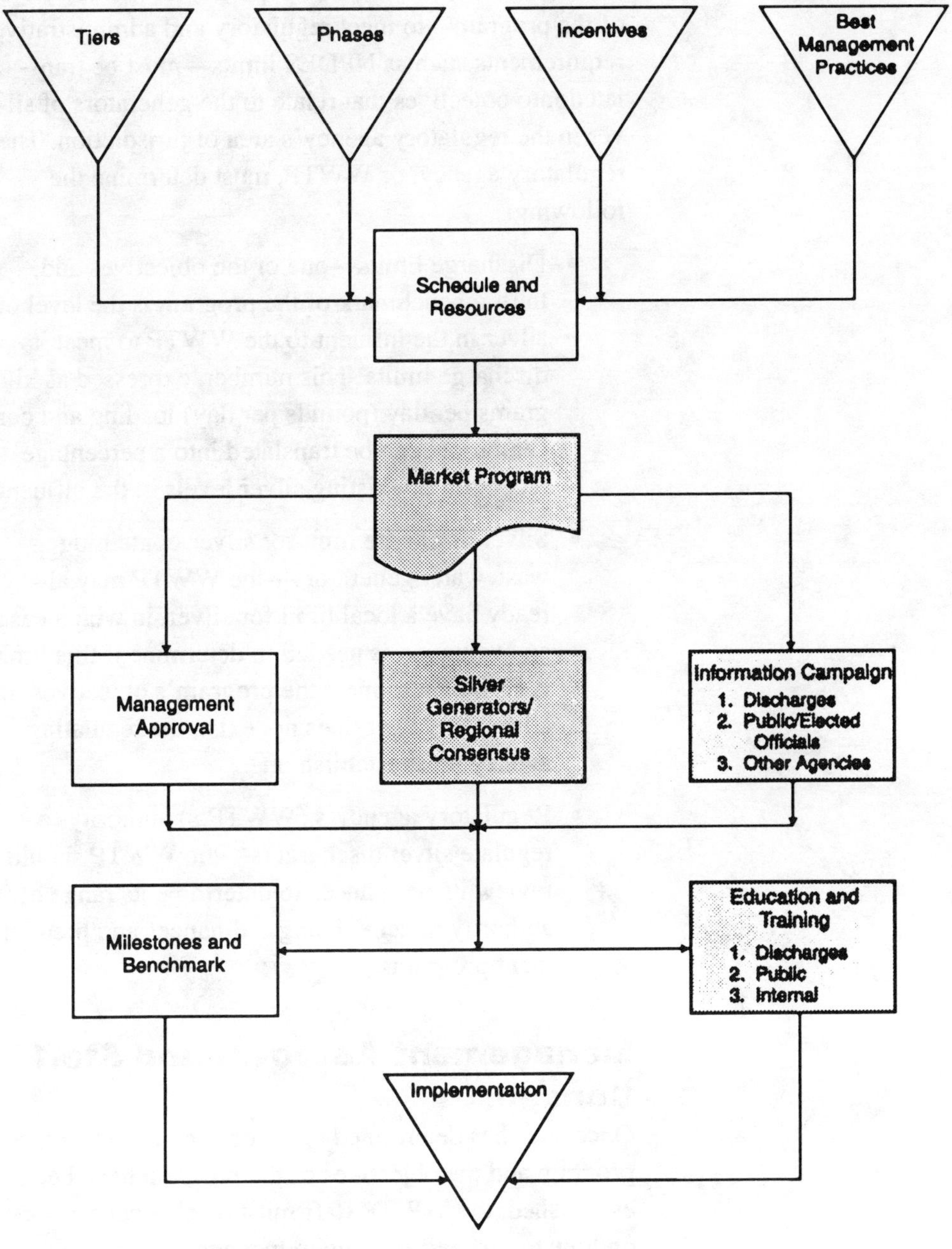

Figure 3 **Silver source control program development, phase II.**

The regulatory basis and the legal boundaries dictate the objectives and goals of the program. The objective of the program—to meet regulatory and administrative requirements such as NPDES limits—must be translated into objectives that relate to the generators of silver in the regulatory agency's area of jurisdiction. The regulatory agency, or WWTP, must determine the following:

- Discharge limits—one of the objectives and, thus, a benchmark of the program is the level of silver in the influent to the WWTP to meet its discharge limits. This number, expressed as kilograms per day (pounds per day) loading and concentration, can be translated into a percentage reduction of existing silver levels in the influent.
- Silver discharge limit for silver-containing wastewater generators—the WWTP may already have a local limit for silver, in which case an evaluation is needed to determine if this limit is sufficient to meet the program's objectives. If a limit for silver does not exist, the regulating agency must establish one.
- Regulatory agency's (WWTP's) authority to regulate silver dischargers—the WWTP should review its ordinances to determine its range of authority under existing ordinances and pretreatment programs.

Management Approval and Staff Consensus

Once staff has determined the need for a silver-control program and the objectives of the program have been established, the WWTP staff must reach a consensus on how to proceed and obtain management's approval. Management and staff consensus at this point in the program are key elements, particularly because the program will require an expenditure of resources and participation of other staff members during the implementation phase.

The following three steps in the development of the program should happen simultaneously and are a collaborative effort of the program stakeholders—the WWTP, silver generators, and other regional, state, or federal regulatory agencies.

Industry/Regional Advisory and Planning Group

Before any further decisions are made regarding the program, the WWTP should set up an advisory group comprising silver dischargers and other regulatory agencies that may affect or be affected by the silver dischargers and the source control program. Local trade associations, national associations, and coalitions can reach a broad membership and bring expertise to the program. Through these groups and members of the advisory group, the WWTP may obtain buy-in for the program. Advice and comments on the program's objectives and practicality should be sought from this group throughout the development of the program. This is particularly important in the case of a silver source control program because it involves mostly small businesses and commercial facilities.

Identifying Sources of Silver

The regulatory agency/WWTP must identify, both qualitatively and quantitatively, the silver dischargers in its area:

- Who are the generators and dischargers? The obvious sources are photoprocessors and X-ray laboratories. Photoprocessors can be either part of a chain, whereby film is developed centrally, or individual photoprocessors who develop film on the premises. X-ray film development takes place in hospitals, imaging centers, and dentists' offices. Less obvious and numerous sources are nondestructive tests of piping and vessels done by metal fabricators.

- How many silver generators are in the WWTP's service area? Once the potential sources are qualified, an estimation of the number of these sources and their sizes should be made. The WWTP should determine if there are large, central photoprocessors, film studios, and large hospitals or medical centers present in its service area.

Information on the types and number of sources can be found in telephone directories, trade associations, chambers of commerce, and other regulatory sources such as health care agencies, agencies regulating hazardous wastes, municipalities, and tax assessors' rolls.

Program

After the preparatory actions identified above have been completed satisfactorily, the actual development of the silver source control program begins. The following steps should be taken with the contribution of the advisory group and the consensus of an internal team.

Evaluation of Silver Dischargers. The evaluation of silver dischargers entails obtaining enough detailed data to establish the scope and range of the program. This step is also meant to allow the program team to become knowledgeable with the issues faced by dischargers. The following actions are suggested to obtain the information to evaluate the kinds of silver sources in the WWTP's service area:

- Conduct a survey of dischargers and compile the data obtained from the survey to determine any trends or special issues. The survey, which can be carried out with the help of the advisory members, should be simple, with only a few basic questions to establish the main data points. A survey only provides trends and qualitative data because, unless it is mandatory, only a limited number of dischargers will return the questionnaires.

- Conduct inspections of selected sites to obtain information in the field and observe issues noted in the surveys.
- Conduct literature searches to determine the number and location of sources. Literature sources include the telephone directories, data from other agencies that in any way license or permit the photoprocessing facilities, and other users of photoprocessing.

Silver Dischargers Assessment. With the help of surveys, site inspections, experts in the advisory group, and suppliers of the developing materials, an assessment of technology, substances, and materials involved in photoprocessing is necessary to obtain the following data:

- Types of photoprocessing and developing technologies and the types of discharges expected from each (such as color, black and white, or X-ray).
- State-of-the-art technologies for the pretreatment of photoprocessing discharges and silver recovery. Based on the surveys and inspection information, evaluate the level at which pretreatment technology and silver-recovery practices are being applied by the silver dischargers in the WWTP's service area.
- Pollution prevention practices and technologies such as material substitution or process modifications to reduce the generation of silver in photofinishing and X-ray developing processes.

Survey of Similar Programs. Several sanitation agencies and WWTPs have already developed and implemented programs to control the discharge of silver to their facilities. A survey of these programs provides important background information on what is effective and on the issues faced by both agencies and dischargers.

Policies and Tools of Implementation

The first phase of the program development, as described above, establishes a database of information on dischargers, photoprocessing technologies, pretreatment, and pollution prevention practices. Based on these data, the program enters the strategy phase of development, where policies and tools of implementation are evaluated and incorporated in the program.

The following are some basic policies that can be implemented:

- Every photoprocessing lab should recover all silver-bearing waste streams;
- Silver-recovery units should be designed and maintained to meet or exceed discharge limits for silver;
- Small labs should use washless/plumbingless systems to minimize water consumption;
- All dischargers should adopt and implement best management practices (BMP); and
- Large photoprocessors (wholesale) should evaluate additional options to pretreat discharge, reduce water consumption, reuse washwater, reduce chemicals, and recover silver.

Additional policies tailored to local conditions can be developed by individual WWTPs.

Tools of implementation include:

- Significant industrial user permits for large facilities;
- Special permitting for small facilities;
- Certification (silver-recovery system installation, waste hauling, and zero discharge, for example);
- Hauling manifests for recovered silver and other developing substances and shipping and purchase records of discharges taken by a silver recycler;
- Monitoring;
- Inspections;
- Local limits;
- Ordinances;
- Enforcement actions; and
- Pollution prevention plans (waste management plans).

Best Management Practices for Silver Dischargers

A silver-discharge-control program often deals with small, commercial dischargers such as small photoprocessing shops, dental offices, or laboratory facilities. Because of the size of the operations, most of these facilities have little, if any, expertise in running a full pretreatment or silver-recovery system. Commercial facilities and other small businesses would be better served by managing their silver discharges to reduce the need for pretreatment or silver-recovery systems. Therefore, BMPs are a key part of the silver-discharge-control program to be adopted and implemented by small facilities. Best management practices should rely heavily on pollution prevention methods and technologies and be developed in collaboration with the advisory group and representatives of the industry.

Implementation Strategy

The program, depending on the size and range of dischargers, will have to be implemented in phases and with a tiered approach. Tiers can be based either on the size, type, or both size and type of dischargers. The first tier implemented can be considered the pilot phase of the program.

In addition, the strategy of implementation must include incentives for dischargers to reduce silver discharges. Examples of incentives that can be applied are

- Reclassifying permit status or level—if a large discharger implements long-term pollution prevention practices and BMPs to reduce silver generation and installs silver-recovery units, the WWTP may be able to reclassify the discharger from an SIU to non-SIU, if allowed under the federal categorical requirements. Hospital X-ray laboratories are one possible example.
- Providing economic incentives—the use of silver surcharges, or ammonia surcharges, can be an incentive to improve silver recovery or, in the case of small dischargers, to ship all silver-bearing wastes to a central recycler.
- Reducing enforcement actions for dischargers that install recovery units.
- Offering simplified certification that eliminates permit fees and surcharges.
- Providing zero discharge or recovery unit certification or reducing inspection frequency and fees for such facilities.

Schedule and Resources Requirements

As soon as the scope of the program and the tools of implementation have been established, a schedule for implementing the phases of the program and resources requirements should be evaluated (see Figure 3).

Program Marketing

Marketing the program internally and externally is one of the most important phases of the program and typically the first phase of implementation. This phase has three steps:

- Obtaining management approval—management must approve the program, schedule, and resources required.
- Informing the region and dischargers of the program and soliciting comments and suggestions—at this point, comments and suggestions should be minimal because the program has been developed with continuous input from the advisory group.
- Developing and conducting an informational and educational campaign for dischargers, the community, and, if required, elected officials.

Milestones and Benchmarks to Evaluate Program Progress and Success

To mark progress toward meeting the objectives, milestones and benchmarks must be set. Measurements to determine whether the milestones and objectives are being met should be conducted on several levels: discharger's facility; discharger's effluent; influent to the WWTP; and WWTP effluent (and/or sludge). These benchmarks should be placed alongside a set schedule. The following are some examples of benchmarks and milestones that can be used to measure progress:

- Percentage (number) of facilities that have installed silver-recovery units,
- Percentage of facilities that recover all of the silver either on site or by shipping to recyclers,
- Percentage of facilities that have adopted BMPs,
- Percentage of facilities that are not in violation of local limits, and
- Wastewater treatment plant influent and effluent quality.

Measurement tools include periodic inspections, effluent and influent samples, and discharger surveys.

Education and Training

Many dischargers, especially small commercial facilities, are unaware of regulatory requirements and the effect their activities have on the WWTP and the environment. Consequently, they may operate their facilities without any provisions to reduce contaminants such as silver discharged to the sewers. Therefore, education and training must be key components of any program. Education and training involve both internal staff and dischargers. Educational tools include workshops, seminars, visits by inspectors with BMP educational material, and newsletters. Key disseminators of information may be local trade associations and photoprocessors' coalitions. Another key educator may be inspectors who come in contact with dischargers on a daily basis. Inspectors can provide educational materials, discuss BMPs, and point out the areas in which the photoprocessors can improve operations and maintenance to reduce discharges.

Example Program

Figure 4 provides an example of a tiered program for silver dischargers. This program for photoprocessors considers two tiers: Tier 1 is for large, central photoprocessors that receive film for developing from small outlets, and Tier 2 is for small, commercial photography laboratories such as 1-hour minilabs. The type of permit and level of administration, such as inspection and monitoring of discharges, is tailored to the size of the facility. Tier 1 photoprocessors are likely subject to an SIU-type permit, while Tier 2 facilities would likely be governed by a special, simplified permitting process requiring no more than yearly monitoring (inspection).

The SIU permits, special permits, and certification incorporated into the example program all include provisions for inspection, monitoring, and, in cases of zero discharge facilities, recordkeeping. Permitting and certification are used as incentives for small, commercial facilities to reduce waste and attain zero discharge through methods such as silver recycling. (Although a silver recycler also produces discharges to the sewer, it typically is better equipped to treat waste and monitor discharges. Also, the central recycling facility would be regulated under an SIU permit and, therefore, receive better control. This is more practical and easier to administer for an agency than several individual commercial facilities.)

Incentives to reduce waste and encourage full silver recovery may include the level of equipment required, fees to the WWTP, and level of inspections and monitoring. A zero discharge facility, for example, would not have to pay surcharges for silver, install silver-recovery equipment, or be subject to discharge monitoring. A certification and recordkeeping system with a biannual inspection may represent lower costs to the discharger than the cost of a discharge permit.

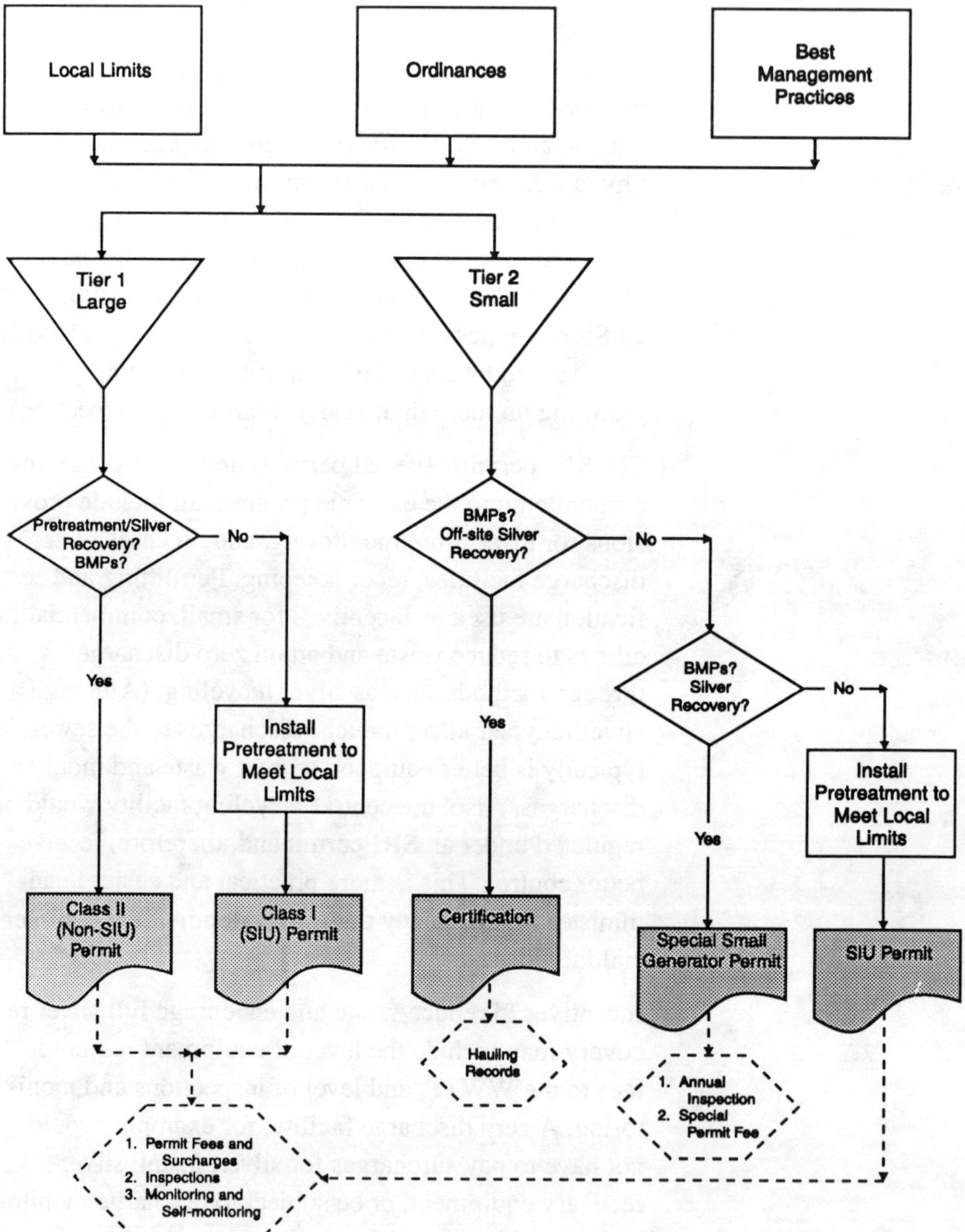

Figure 4 **Silver source control program for photoprocessors.**

CASE STUDY—CITY OF BOCA RATON

Chris Ferrigan

The City of Boca Raton Wastewater Treatment Plant, Boca Raton, Florida, has an aggressive and effective source control program. The city regularly monitors industries or businesses that generate photographic or X-ray waste. The city especially has a concern during the months of October to April, when the potential for conventional and metal pollutants is highest. Some businesses use the sewerage system as their only means of disposal for these pollutants.

The city has focused not only on possible silver problems but on other metals and conventional pollutants that most cities, utilities, or inspectors may not be aware of as potential pollutants of concern. The City of Boca Raton's metals-monitoring program has been successfully operated since 1987.

Treatment of silver from fixers has been accomplished using metallic replacement, electrolysis, and ion exchange units (see Silver-recovery Technologies section). It may be possible to meet the local silver parameters if the silver-recovery units (SRUs) are operated and maintained on a regular basis. One of the problems with the SRUs is that they are not flow-proportioned and are often hooked up and left running until the filters or lines clog and/or silver or iron deposits clog the drains. The SRUs are then unable to meet local silver limits consistently. These units are designed to reduce or eliminate silver but not other pollutants of concern. Some users may increase the rinse water for purposes of dilution. Dilution, as stated in 40 CFR, is not an acceptable means of treatment.

To determine whether the discharge meets local limits, a sample must be taken at the end of the process or from the discharge that meets the drain. This can be accomplished by one or more lines (fixer and rinse water and a second line from the developer). If the process is not running at the time of collection, sample blanks must be inserted to get a representative sample. The sample technician should be sure the rinse water has not been turned up for dilution. After the grab samples have been collected (following proper collection procedures from 40 CFR Part 136, chain of custody, for example), all related correspondence, including telephone logs (Figure 5) and pictures showing discharge lines and possible poor housekeeping are documented. Standard procedures established in the local Enforcement Response Plan must be followed.

After each photographic and X-ray inspection, the trained inspector should provide a list of all certified and legal companies (as well as state satellite hazardous waste drop-off sites) that offer proper disposal services of spent waste including sludge from ion exchange. Waste-minimization and reduction practices should be encouraged whenever possible. There are new SRUs on the market that regulate and reuse rinse water, fixer, and developer until the solution is spent.

Procedures used by the City of Boca Raton to establish a program to monitor SRUs are as follows:

1. Obtain a list from the local occupational or business-licensing department of all industries of concern, such as photography laboratories, medical clinics, hospitals, doctor and dentist offices, and printers and publishers.

2. Send out "certified" letters to all users on rules and regulations, with time frame to comply (see Figure 6).

3. Begin inspections (see Figure 7).

City of Boca Raton Public Utilities

Telephone Log

Date and Time		Comments
12/09/93	12:45 p.m.	Called and talked to owner about response to certified letter. He was not available; left message to return call to city.
12/10/93	9:35 a.m.	Owner called and said he plans on using his specialists and continuing his process. I notified him that we would need to put him on a schedule and he would furnish us test results at his expense.
01/06/94	4:00 p.m.	Owner called and claims that the city officials did not take a representative sample according to his manager. I explained to him that we did and that we could take a sample from any drain that discharged to the city sewer system. I also told him that we collected a sample from his overflow line (from ion exchange) that also was discharging to city sewer. He said he would be in tomorrow to correct that problem.
01/07/94	1:00 p.m.	Owner called and left message that he had fixed any loose lines.
01/11/94	9:00 a.m.	Checked to see if any lines had been removed from drain; we could see no changes made at all. Overflow line still discharging to city drain.
01/11/94	9:30 a.m.	Called U.S. EPA Region IV and asked if I could issue penalty to facility under SNC. He said he thought so, but would check with Chief of Pretreatment and get back to me.
01/11/94	11:40 a.m.	U.S. EPA Atlanta called back and said he talked it over with the chief, who agreed that we could issue penalty under SNC rules 403.8(F) and (H) plus charge for analysis expenses. He also said that we must publish them in the newspaper under 40 CFR 403.11(B). I also told U.S. EPA that the owner plans to call U.S. EPA. They said, "He could call, we will back the city all the way."

Figure 5 **Sample telephone log.**

Date

Address

To Whom It May Concern:

Subject: Hazardous Waste

The purpose of this letter is to inform doctors, dentists, medical clinics, printing shops, and photo shops discharging photograph/X-ray processing that they must either treat on site or contain the waste in an approved container. The waste then must be held and hauled by a properly licensed waste hauler.

If you choose to treat your waste by installing a primary recovery unit (metallic, electrolytic, or ion exchange), you must prove to the city through analysis, at your expense, that you meet the local limits. The city will determine the frequency and require a city official be present at the time of collection and make certain that the sample is a representative sample. The city has the authority as required by CFR 403.8(f)(I)(v) to carry out all inspections, surveillance, and monitoring procedures necessary to determine compliance or noncompliance.

Once an accurate sample has been obtained, the sample should be properly preserved and promptly delivered to a certified laboratory. City personnel will document the methods used to collect the sample as well as the chain possession of the sample from collection to analysis. All analysis must follow methods from 40 CFR part 136.

The waste from the abovementioned process must not be discharged into the sanitary sewer because it has the potential to cause disruption to the wastewater treatment plant or pass through the plant and contaminate the environment.

City Code, Section 17-103, states

(h) Solid or viscous substance which may cause obstruction to the flow in sewers or interference with the proper operation of the city wastewater facilities is prohibited.

(i) Any waters or waste containing toxic, poisonous or radioactive substance in concentrations that may constitute or create a public nuisance or a hazard to humans or animals, or may interfere with any wastewater treatment process or the city wastewater facilities, is prohibited.

Any person who violates any of the provisions of this article shall be liable to the city for all costs and damages incurred by the city as a proximate result of such violations plus a fine up to $1 000 per day.

If necessary, water and wastewater service would be discontinued to ensure compliance with the city code.

In order to avoid this problem, the City of Boca Raton Public Utilities Department intends to inspect on an unannounced basis and, when necessary, take appropriate corrective action to ensure compliance.

We recommend that your facility

a. Keep records or receipts on your firm's cleaning or recovery of waste and

b. Designate one person to monitor this activity and check on a routine basis.

Should you have any questions, please contact the Industrial Wastewater Inspector at 555-1234.

For those establishments conforming with regulations, your continued cooperation is appreciated.

Sincerely,

Director, Public Utilities

Figure 6 Sample notification letter.

City of Boca Raton
Public Utilities Quality Control

Date ______________________ Time ______________________

Name of Business ______________________

Address ______________________

Owner ______________________

Contact Person ______________________

Reason for Visit

First Visit ____________

Follow-up ____________

Grease Trap ____________

Demand, Reason ______________________

____

____

Other ______________________

____

No Contact Made ____________

Status

In Compliance ____________ Noncompliance ____________

Other ____________

Comments ______________________

____

Action Taken ______________________

____

Owner Must Isolate and Correct Problem(s)

Immediately ____________ Within 24 Hrs ____________ Within 3 Days ____________

Other ______________________

Will Return to Check Compliance on Date ______________________

Signature
City Official ____________

Signature
Owner/Representative ____________

original - City of Boca Raton
copy - Owner/Representative

Figure 7 **Inspection report.**

4. Collect samples and have them analyzed. Take photos, if necessary, to show discharge lines to drain or poor housekeeping.
5. If a violation has occurred, send a notice of violation to the violator with a time frame specified to isolate and correct the violation (Figure 8).
6. If the business fails to comply within the prescribed time frame, follow standard procedures set up in the local Enforcement Response Plan, including a legal review of procedures (Figure 9).

A city's source control program decisions come down to how clean the effluent must be immediately and in the future. According to U.S. EPA, the Clean Water Act, the General Pretreatment Regulations, and the City of Boca Raton sewer-use ordinance, the use of the public sewer system is a privilege, not a right, and all users are subject to the same rules and regulations. Pollution prevention begins at the local level.

CASE STUDY—CITY OF SAN BERNARDINO WATER DEPARTMENT PHOTOGRAPHIC WASTE

James H. Watson

During 1981, the City of San Bernardino Water Department developed and implemented its pretreatment program for the City of San Bernardino Water Reclamation Facility. Like most programs during the 1980s, large dischargers were targeted, permitted, and monitored for compliance. By the 1983 deadline, 21 industries had been identified as subject to pretreatment regulations for discharging a host of various pollutants, including toxic heavy metals, total toxic organics, and conventional constituents.

November 18, 1993 CERTIFIED MAIL

Boca Raton, FL 33431

Subject: Violation of Wastewater Discharge

To Whom It May Concern:

On October 27, 1993, a sample of wastewater discharge was collected by (Name), Industrial Wastewater Inspector for the city, and analyzed. The results show that several parameters have exceeded the maximum allowable limits.

Parameter	Results Reported	Maximum Limit
BOD	3960 mg/L	400 mg/L
COD	38261 mg/L	800 mg/L
Cyanide, total	38.4 mg/L	0.25 mg/L
Hydrogen sulfide	3922 mg/L	5.0 mg/L
Iron, total	1828 mg/L	10.0 mg/L
Phenols	25.8 mg/L	0.20 mg/L
Silver, total	280.0 mg/L	0.35 mg/L
Solids, total	57126 mg/L	2400 mg/L
Solids, total dissolved	56918 mg/L	2400 mg/L
Total Kjeldahl nitrogen	10933 mg/L	35.0 mg/L

Please respond in writing with action taken to locate and correct this problem within ten (10) days after receipt of this letter. At that time, you need to retest your facility's wastewater discharge for the parameters of concern and forward the results to my office.

If you continue to exceed the maximum allowable limits, you will be in Significant Noncompliance and subject to a penalty of $1 000 per day.

If you have any questions, please call 555-1234. Thank you for your cooperation.

Sincerely,

(Name)

(Title)

Attachments

Figure 8 Notification of violation, including laboratory results.

Water
Soil
Air
Analysis and Consulting

CITY 00005375

Attn:

Page 1
15 Nov 1993
Report T-3-10-142-01
Lab ID. 84340,E84217,84271,E84060,82223,E821

City of Boca Raton
1501 Glades Road
Boca Raton, FL 33432

Sample Description:

Company Name

Sample ID: Photo
Collected: 10/27/93
Received: 10/27/93
Collected By:

Parameter	Result	Units	Method	Det. Limit	Extracted	Analyzed	Analyst
Biochemical oxygen demand	3960	mg/L	405.1	2.0		10/29/93	ML
Chemical oxygen demand	38261	mg/L	410.4	5.0		11/01/93	ML
Cyanide, total	38.4	mg/L	335.3	0.010		11/09/93	DD
Hydrogen sulfide	3922	mg/L	367.1	0.1		11/04/93	DD
Iron, total	1828	mg/L	236.1	0.015		11/15/93	PR
Phenols (4-AAP)	25.8	mg/L	402.2	0.1		11/10/93	DD
Silver, total	280.0	mg/L	272.2			11/12/93	PR
Solids, total	57126	mg/L	160.3	2.0		11/07/93	CL
Solids, total dissolved	56918	mg/L	160.1	2.0		11/07/93	CL
Solids, total suspended	208	mg/L	160.2	2.0		10/30/93	PR
Total Kjeldahl nitrogen	10933	mg/L	351.2	0.03		11/09/93	CL
pH (lab)	7.06		150.1	0.1		10/29/93	DS

**** BDL indicates analyte is below detectable levels.
All analysis performed by EPA, ASTM, or Standard Methods.

Laboratory Director

Figure 8 Notification of violation, including laboratory results (continued).

January 27, 1994 CERTIFIED MAIL

Boca Raton, FL 33421

Subject: Violation of Wastewater Discharge

Dear (Name):

Last year on October 27, 1993, a sample of wastewater discharge was analyzed and the results showed several parameters exceeded the maximum allowable lists. A certified letter was sent asking you to take action to locate and correct the problem within ten (10) days after receipt of the letter. On December 9, 1993, a phone call was made to remind you that you still needed to reply back to the city. You stated that you were going to continue using (Name) products and felt confident everything would be within the allowable limits and that your letter would be forthcoming. To date you have failed to respond to the city.

On January 4, 1994, a sample of wastewater discharge was collected by (Name), Industrial Wastewater Inspector for the City, and analyzed. The results show that several parameters have exceeded the maximum allowable limits.

Parameter	Results Reported	Max. Limit
BOD	2750 mg/L	400 mg/L
COD	48575 mg/L	800 mg/L
Hydrogen sulfide	9500 mg/L	5.0 mg/L
Iron, total	8420 mg/L	10.00 mg/L
Silver, total	14.4 mg/L	0.35 mg/L
Solids, total	126984 mg/L	2400 mg/L
Solids, total dissolved	120664 mg/L	2400 mg/L
Solids, total suspended	6320 mg/L	400 mg/L
Total Kjeldahl nitrogen	22480 mg/L	35.0 mg/L

Your business is in Significant Noncompliance (SNC) due to violation of the following criteria from 40 CFR 403.8 vii (Code of Federal Regulation):

(F) Failure to provide, within 30 days after the due date, required reports such as baseline monitoring reports, 90-day compliance reports, periodic self-monitoring reports, and reports on compliance with compliance schedules;

(H) Any other violation or group of violations which the Control Authority determines will adversely affect the operation or implementation of the local pretreatment program.

This is to advise you that your business is subject to a penalty of $500 plus lab expenses of $283.50.* Please make your check payable to the City of Boca Raton Public Utilities for a total of $783.50 and mail it to Industrial Pretreatment Division, 201 West Palmetto Park Road, Boca Raton, FL 33432, no later than ten (10) days after receipt of this letter.

In addition, please respond in writing within ten (10) days of this letter with action taken to locate and correct these violations. As soon as the violations have been corrected, you will need to retest your facility's wastewater discharge for the parameters of concern and forward the results to the address noted in the paragraph above. A city official must be present during sampling.

If you have any questions, please contact (Name), Industrial Waste Coordinator, at 555-1234.

Sincerely,
(Name)
(Title)
*See attached invoices

Figure 9 Notification of significant noncompliance, including laboratory results.

Water
Soil
Air
Analysis and Consulting

CITY 00005375

Attn:

Page 1
12 Jan 1994
Report T4-01-018-01
Lab ID. 84340,E84217,84271,E84060,82223,E821

City of Boca Raton
1501 Glades Road
Boca Raton, FL 33432

Sample Description:

Company Name

Sample ID: Photo
Collected: 01/04/94
Received: 01/05/94
Collected By:

Parameter	Result	Units	Method	Det. Limit	Extracted	Analyzed	Analyst
Biochemical oxygen demand	2750	mg/L	405.1	5.0		01/05/94	CL
Chemical oxygen demand	48575	mg/L	410.4	5.0		01/05/94	DS
Cyanate*	BDL	mg/L	412.K	100		01/07/94	DD
Cyanide, total*	BDL	mg/L	335.3	100		01/07/94	DD
Hydrogen sulfide	9500	mg/L	367.1	0.1		01/06/94	TBY
Iron, total	8420	mg/L	236.1	1.0		01/06/94	DP
Silver, total	14.4	mg/L	272.2	0.1		01/06/94	DP
Solids, total	126984	mg/L	160.3	2.0		01/07/94	CL
Solids, total dissolved	120664	mg/L	160.1	2.0		01/06/94	CL
Solids, total suspended	6320	mg/L	160.2	2.0		01/05/94	DS
Total Kjeldahl nitrogen	224800	mg/L	351.2	0.05		01/05/94	DS
pH (lab)	7.3		150.1	0.1		01/05/94	DS

* Had to go to a higher detection limit because of interference of silver and iron. Results were probably higher than local limit. Normal detection limit is 0.010. See 10-27-93 Report.

**** BDL indicates analyte is below detectable levels.

All analysis performed by EPA, ASTM, or Standard Methods.

Laboratory Director

Figure 9 Notification of significant noncompliance, including laboratory results (continued).

Throughout the next decade, the city tightened its permit allowance and increased enforcement practices. During this period, they continued to monitor large industry, realizing there were many small establishments discharging a variety of undetected elements. The city received numerous phone calls from photography laboratories or similar establishments asking what to do with their waste. This area had not yet been considered, so the decision was made to permit all photoprocessing waste facilities. By placing them on permit, the city could regulate their discharges and require pretreatment in the form of silver-recovery units on the waste streams. By 1990, all hospitals and large laboratories were in full compliance with discharge of silver-bearing waste to the sanitary sewer system. A year later, all 1-hour photography shops and similar establishments were in compliance.

Influent silver was reduced from an average of 0.2 mg/L (200 ppb) in 1983 to 0.01 mg/L (10 ppb) in 1991; however, the Santa Ana Regional Water Quality Control Board (RWQCB) revised the San Bernardino facility NPDES permit's allowable silver discharge limit from 0.05 mg/L (50 ppb) down to 0.006 mg/L (6 ppb), a tenfold reduction. It became necessary to identify all sources of silver in the sanitary sewer system, including small businesses that only produced approximately 19 L/month (5 gal/month) waste, and, if possible, eliminate their discharges.

A survey was sent out to all doctors, dentists, chiropractors, veterinarians, and a few home portrait businesses. It became clear that these unregulated businesses were discharging their wastes into the sewer system on a sporadic basis, though how much and at what strength was not as clear. Inspections conducted at the surveyed premises and subsequent sample analyses indicated silver in these businesses' waste up to 4 000 mg/L and a chemical oxygen demand of 40 000+ mg/L, which was going directly into the sewer system.

Waste Minimization

The survey had identified approximately 60 sources of silver-bearing photoprocessing waste being discharged into the sewer system in unknown amounts. The city recognized that these premises were small generators, some as small as 1 L (1 qt) per 6 months, that had no avenue for recycling. Assistance, rather than enforcement, was needed to resolve the issue. A pilot program with a simple format was initiated to eliminate the discharge of silver-bearing photoprocessing waste.

Businesses discharging into the sewer system were asked instead to pour the photoprocessing waste into buckets supplied by the city and approved by the Department of Transportation. Once filled, the buckets were brought to the WWTP for correct disposal. The pilot program received approximately 2 m^3 (500 gal) of photoprocessing waste in its first year, which would have been disposed of improperly to the sanitary sewer system.

The San Bernardino facility constructed a small waste-receiving station with standard containment for storage and disposal. Disposal was contracted out to a state-approved waste recycler who picks up waste, recycles the silver, and uses the liquid residual for the manufacturing of cement. Thus, full recycling and waste minimization is accomplished.

Results

At a nominal cost, less than $15 000 to set up the program, silver in the influent has nearly been eliminated. Small generators have been put in compliance with local regulations and have saved a significant amount of money. The city has met all of its monthly NPDES reporting requirements for silver during 1993.

CASE STUDY—SILVER REDUCTION PILOT PROGRAM AT THE PALO ALTO REGIONAL WATER QUALITY CONTROL PLANT

Phil L. Bobel, Matt Chadsey, and Anthony M. Hui

The Palo Alto Regional Water Quality Control Plant (RWQCP) discharges treated municipal and industrial wastewater into South San Francisco Bay. The wastewater is a high-quality tertiary effluent that is low in biochemical oxygen demand, total suspended solids, ammonia, and other conventional pollutants. However, further reductions in heavy metals have been mandated because South Bay waters provide limited dilution.

In 1988, RWQCP received a revised NPDES permit to continue discharge contingent on performing several studies to identify the sources of metals and appropriate reduction methods and assist in development of final water-quality-based effluent limits by the next permit reissuance. To comply with the permit requirements, RWQCP undertook seven studies, including a source identification study and a waste-minimization study. The purpose of the source identification study was to identify all significant controllable sources of metals. The subsequent waste-minimization study identified feasible measures to reduce the metals loadings to the treatment plant.

Based on results of these studies, RWQCP decided to take a proactive approach to waste minimization and proposed the initiation of a Silver Reduction Pilot Program to the Regional Water Quality Control Board (Regional Board). A pilot program focusing on one specific metal would allow the RWQCP to consolidate its efforts and establish the technical, institutional, and regulatory measures necessary to promote source reduction. Experience learned from this pilot program could then be applied to reducing other toxic pollutants. The Regional Board concurred with this approach and directed RWQCP to proceed with the pilot program in a June 1990 amendment to RWQCP's NPDES permit.

Treatment Facilities

The Palo Alto RWQCP treats industrial and domestic wastewater from the cities of Palo Alto, Mountain View, East Palo Alto, Los Altos, Los Altos Hills, and Stanford University. The service area covers approximately 15 000 ha (38 000 ac), consisting of low- to high-density residential development with scattered commercial districts and electronics-oriented industries. Total population within the service area is approximately 200 000.

The RWQCP provides primary, secondary, and tertiary treatment for influent wastewater. The plant has a designed average dry weather flow capacity of 1.7 m^3/s (39 mgd). Because of the ongoing drought in California and water conservation measures, influent flows to the plant have declined from an annual average of 1.2 m^3/s (27 mgd) in 1982 to 0.9 m^3/s (20 mgd) in 1991.

A flow schematic of RWQCP facilities is shown in Figure 10. The major unit processes are primary sedimentation, trickling filters, activated sludge for nitrification, filtration, chlorination, and dechlorination. The effluent is discharged through a 518-m (1 700-ft) long discharge channel that travels through adjacent wetlands to the South Bay. Discharge limitations for conventional pollutants are consistently met. Settled solids from the primary sedimentation tanks and waste biological solids from the secondary clarifiers are thickened in the gravity sludge thickeners and then pumped to belt filter presses for dewatering. Dewatered sludge

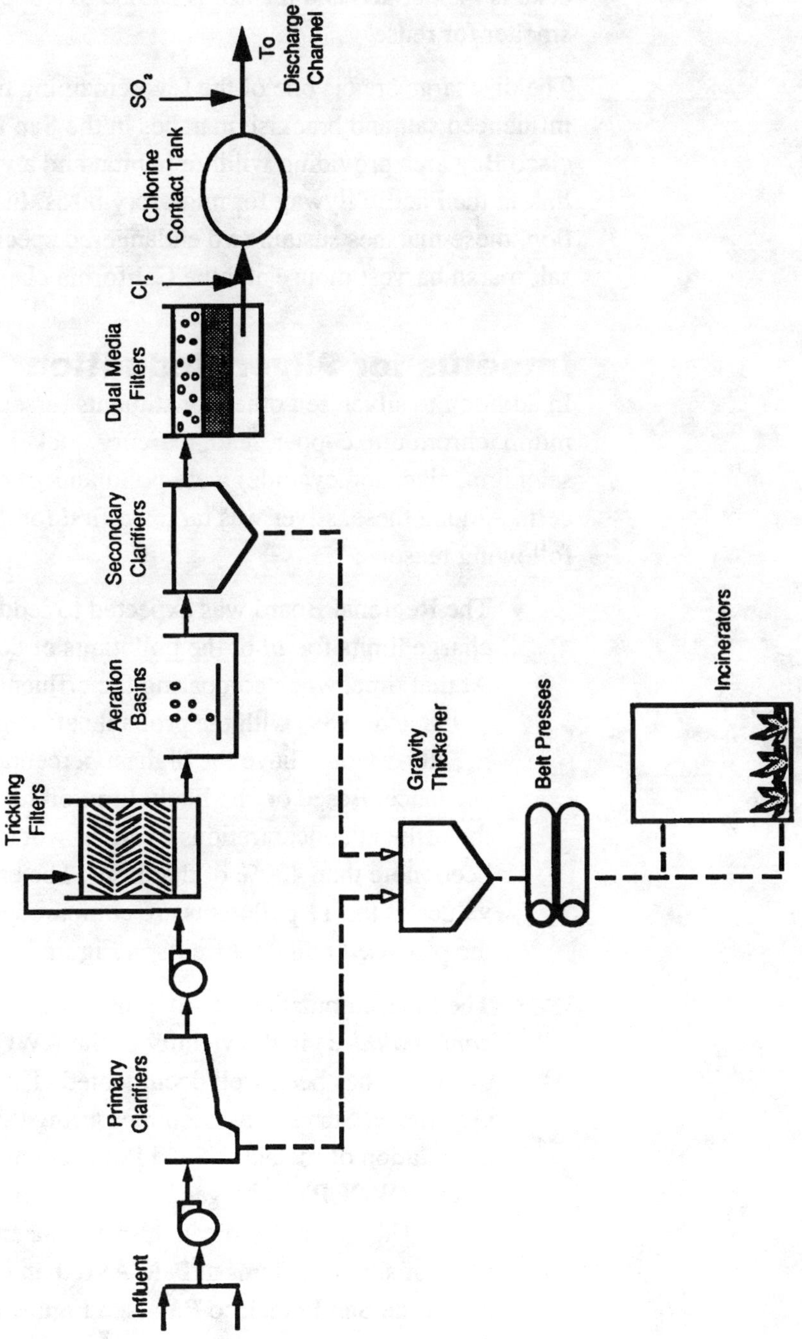

Figure 10 Plant flow schematic.

cake is incinerated and the ash is hauled to a copper smelter for reuse.

The discharge area is one of the few remaining tide-influenced salt and brackish marshes in the San Francisco Bay area providing wildlife habitat and a vital link in the Pacific flyway for migratory birds. In addition, these marshes sustain two endangered species: the salt marsh harvest mouse and the California clapper rail.

Impetus for Silver Reduction

In addition to silver, ten other constituents (arsenic, cadmium, chromium, copper, lead, mercury, nickel, selenium, zinc, and cyanide) were pollutants of concern. Among these, silver was targeted first for the following reasons:

- The Regional Board was expected to reduce discharge limits for all of the pollutants of concern. At that time, when comparing the effluent data (1988 and 1989) with the probable effluent limits, silver would have the highest percentage exceedance. Based on the likely limit of 2.3 μg/L, the effluent concentrations of silver would have been more than 400% of that limit. The effluent values of the 11 pollutants are compared with the proposed pollutant limits in Figure 11.
- The bioaccumulation of silver in clams (*Macoma balthica*) in the vicinity of the RWQCP discharge has been well documented. The U.S. Geological Survey has been measuring the bioaccumulation of metals at Sand Point, 1 km south of the RWQCP discharge channel, for many years. Their results indicate higher concentrations of silver in clams at Palo Alto than other areas of the San Francisco Bay (see Figure 12).
- Among the 11 pollutants of concern, silver is the only precious metal. Because of its value,

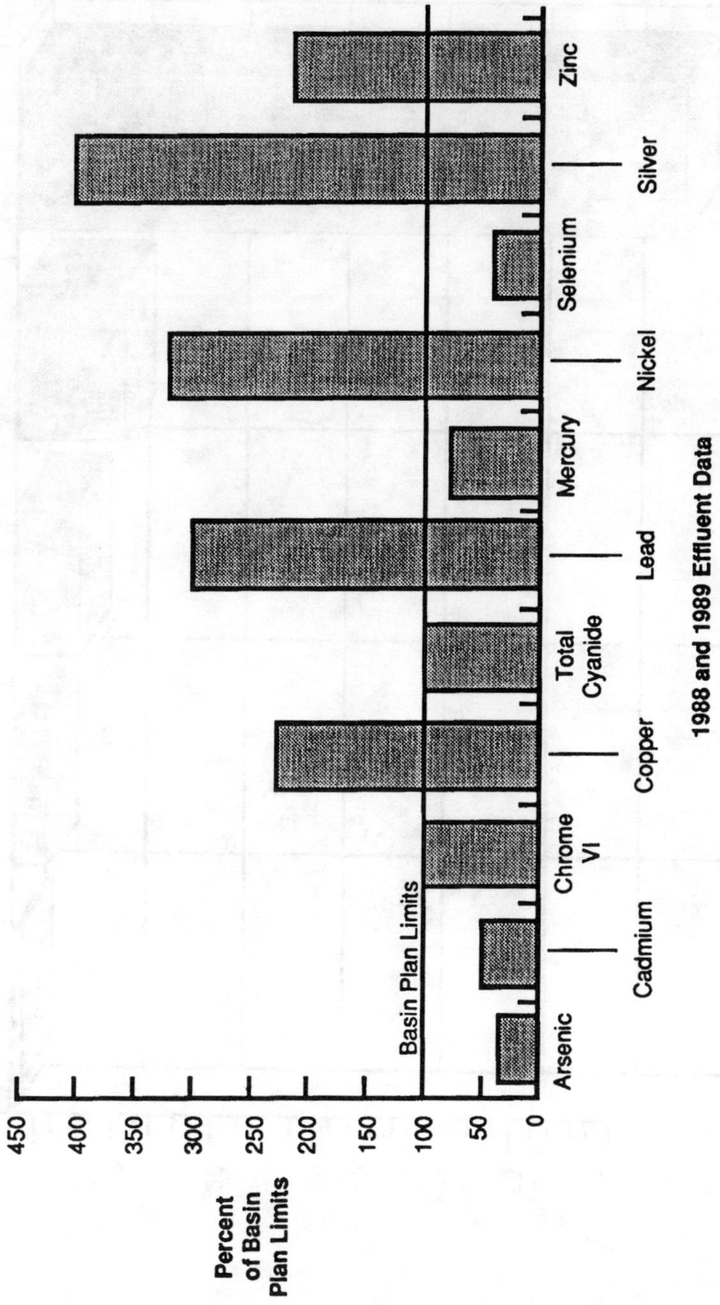

Figure 11 **Comparison of effluent concentrations with basin plan limits.**

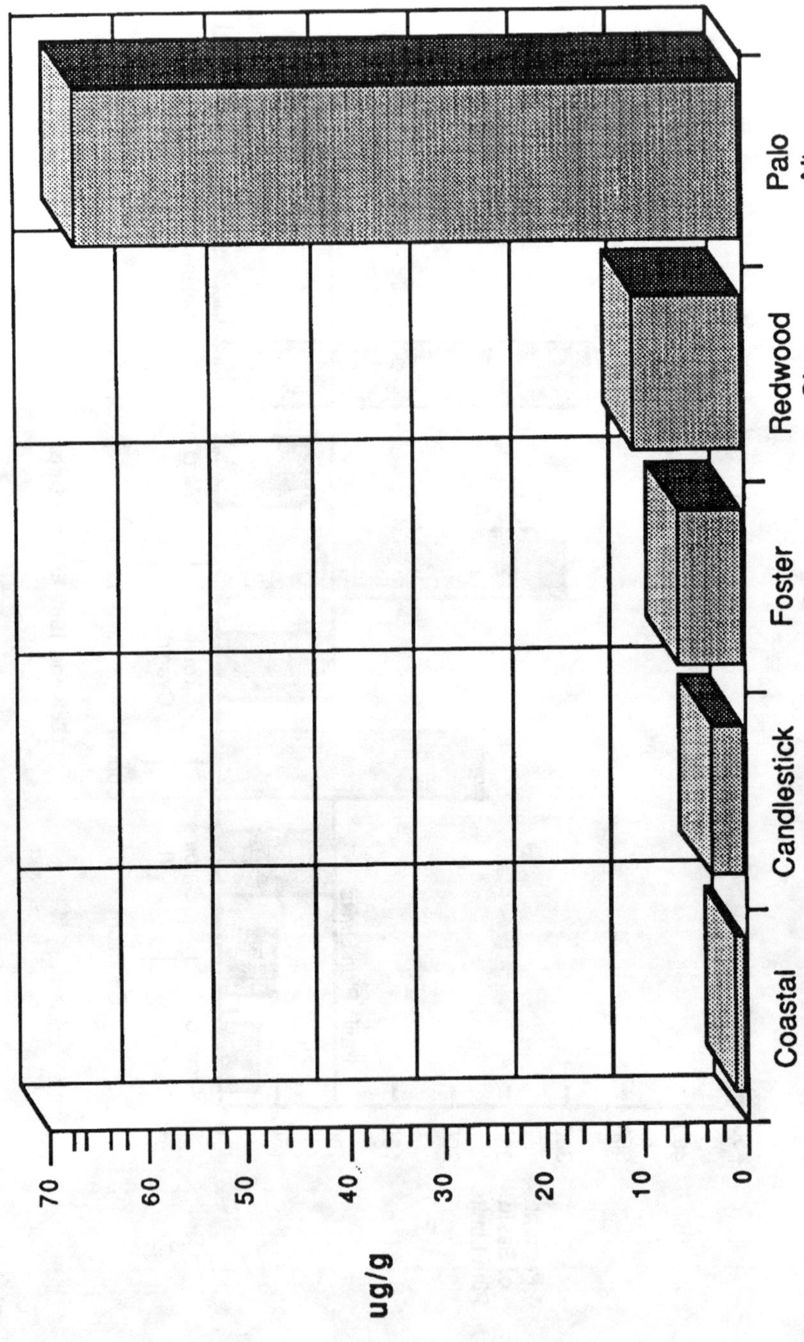

Figure 12 **Silver concentrations in *Macoma balthica* (clams).**

discharges of this metal would have the most significant economic incentive to implement source reduction and waste-minimization measures.

- Most of the silver appeared to be coming from one source—photoprocessing. This commercial category would be the major focus of the pilot program.

Initial Mass Balance

An initial mass balance was performed as part of the 1989 Metals Source Identification Study to quantify contributions of pollutants to the RWQCP. The potential contributors of these pollutants included the regulated industries, commercial businesses, residential sector, and water supply.

Of the four potential sources identified above, only data from regulated industries and the water supply were available. Therefore, sampling of residential and commercial truck lines was performed to quantify their respective contributions. To define these contributions as best as possible, RWQCP coordinated its sampling program with similar programs concurrently performed by the Sunnyvale Wastewater Treatment Plant and San Jose/Santa Clara Water Pollution Control Plant. The average concentrations from these areas were used to calculate the loadings from the commercial businesses and residential contributions.

The results of the sampling effort and mass balance for silver were compiled. The residential, industrial, and commercial contributions were estimated to be 5, 25, and 43% of the plant loadings, respectively. A remaining 27% of the influent silver loadings was unaccounted for. It is postulated that the unaccounted loadings were likely contributed by small commercial businesses and, to a lesser extent, residents. Based on this mass balance effort, it appeared that small businesses contributed from 43 to 70% of the plant's influent silver loading.

Refined Source Identification

The next step in this identification process was to develop a list of business categories that might process photographic materials, X-rays, and photographic films and negatives. Businesses included on the list were small photofinishers, X-ray laboratories, dental offices, hospitals (with photoprocessors), veterinarian offices, art/advertising/graphics firms, chiropractors/orthopedic physician offices, medical clinics, biological research firms, nondestructive testing firms, police stations, publishing companies, and schools/colleges. After establishing a list of businesses in these categories, RWQCP staff conducted a mail survey designed to answer the following questions:

- What facilities process photographic materials?
- What forms of photoprocessing dominate within the service area?
- What volume of silver-rich solutions are produced at each facility?
- What silver-recovery processes are currently used?

A test survey of 12 facilities was performed to evaluate the completeness and effectiveness of the survey forms. After including comments from the test, survey forms were sent to 650 businesses within the service area. A copy of the survey is shown in Figure 13.

More than 70% of the facilities that received the mailing completed and returned the survey, and, of those, 53% indicated that they conducted photoprocessing. A large number of relatively small dental offices and laboratories (47%), printing/publishing (14%), photographic (18%), and art/graphics/advertising

1. DOES YOUR FACILITY PROCESS ANY PHOTOGRAPHIC MATERIALS ON SITE?

YES ________ NO ________

2. BRIEFLY DESCRIBE YOUR OPERATION: ________________________________

____

____

******* IF ANSWER TO QUESTION #1 IS "NO", GO TO #8 *******

3. WHICH OF THE FOLLOWING DOES YOUR FACILITY PROCESS ON SITE?

	COLOR	BLACK & WHITE
FILM	________	________
PAPER	________	________
MOVIE	________	________
X-RAYS	________	________

OTHER: ____

4. WHAT IS THE TOTAL VOLUME OF FIXER AND BLEACH-FIX USED PER MONTH?

______ < 5 GALLONS ______ 26–100 GALLONS ______ > 500 GALLONS

______ 6–25 GALLONS ______ 101–500 GALLONS AMOUNT: ________

5. DOES YOUR FACILITY REMOVE SILVER FROM THESE SOLUTIONS ON SITE?

______ FIX ______ WASHWATER

______ BLEACH-FIX

RECOVER SYSTEM INFORMATION (IF APPLICABLE):

MANUFACTURER: ________________________________

MAKE & MODEL: ________________________________

6. DOES YOUR FACILITY SEND EITHER OF THESE MATERIALS OFF SITE FOR SILVER RECOVERY?

______ SPENT CARTRIDGES ______ SPENT SOLUTIONS

7. COMMENTS/QUESTIONS: ________________________________

____

____

8. FACILITY CONTACT INFORMATION:

BUSINESS NAME: ________________________________

ADDRESS: ________________ TELEPHONE: ________________

CONTACT NAME: ________________ TITLE: ________________

PLEASE RETURN TO RWQCP IN THE ENCLOSED ENVELOPE WITHIN TEN DAYS

REGIONAL WATER QUALITY CONTROL PLANT

2501 Embarcadaro Way - Palo Alto, CA 94303

SERVING THE COMMUNITIES OF EAST PALO ALTO - LOS ALTOS - LOS ALTOS HILLS - MOUNTAIN VIEW - PALO ALTO - STANFORD

Figure 13 Facility survey of photographic activities.

(18%) businesses, and medical facilities (3%) indicated that they processed photographic materials. A survey of fixer use in facilities processing photographic material indicated that 80% use less than 0.019 m^3/month (5 gal/month), 10% use 0.02 to 0.09 m^3/month (6 to 25 gal/month), 7% use 0.1 to 0.38 m^3/month (26 to 100 gal/month), 2% use 0.39 to 1.9 m^3/month (101 to 500 gal/month), and 1% use more than 1.9 m^3/month (500 gal/month).

Overall Program Approach

As a prelude to developing the pilot elements, the RWQCP staff adopted the following approach for the Silver Reduction Pilot Program.

Maximize Public Participation. The cornerstone of this program would be to increase public participation. As many groups and viewpoints as possible would be included in developing and implementing this program. Three reasons necessitate this precept:

- Good, innovative ideas are needed in a pilot program. Businesses know their operations best; environmental groups also have fresh ideas. All viewpoints can be helpful.
- A higher degree of public acceptance can be achieved by hearing each interest group's position and attempting to address these concerns.
- Time-consuming debates and delay can be avoided at the political level by incorporating public participation early in the program development process and creating a program that most everyone can live with.

Other Factors. The program would be referred to as a *pilot.* Because of the innovative nature of the program, it was not certain that all aspects of the program would work. Some of the program components would be experimental and perhaps have to be changed or dropped.

While modifications were inevitable, RWQCP did not want to have to significantly modify the program in the future and ask for additional reductions. To prevent against a "moving target," the lowest likely effluent limit would be used to design the program.

The RWQCP would establish a program that sought reductions from the industrial, commercial, and residential sectors. This approach would avoid the criticism that attention was unfairly focused on one specific group.

On the other hand, RWQCP would not emphasize reductions from a group that did not commensurate with its contributions and its relative cost for reducing silver. Thus, the commercial sector, contributing 43 to 75% of the influent silver, would get the most attention.

The RWQCP would use the ordinance process of all of its member agencies to impose the new requirements. This step ensured that everyone, including public officials, were focusing on the problem, allowed everyone to be heard publicly, and avoided any confusion on what the enforceable requirements would be. Some of the new provisions did not have to be included in the ordinance itself; however, they were included for emphasis and clarity.

Personal contacts, mailings, workshops, and media coverage would be used to provide information and achieve awareness of the need for and details about the program. The communications also reinforced the concept that RWQCP staff would help the businesses through the program.

The RWQCP would commit to working to resolve conflicts among agency requirements that make it difficult for the businesses to comply with the new requirements. While not always successful, the recognition of the problems and the time spent on them improved credibility and the relationship with the businesses substantially. A number of the efforts were successful.

Program Implementation

Having defined the guidelines of the program, specific implementation steps were developed into a cohesive and coordinated Silver Reduction Pilot Program. The key components of this program (public participation, ordinance revisions, industrial reduction activities, commercial reduction activities, and residential reduction activities) are discussed below.

Public Participation. The first step was the formation of the Metals Advisory Group to advise RWQCP on program components. The Metals Advisory Group comprised approximately 20 individuals representing environmental groups, individual businesses (small and large), trade associations, regulatory agencies, RWQCP staff, and concerned citizens. To make the meetings more productive, the group met only when sufficient products or program ideas were available for review. Materials were sent to members in advance for review, and comments were recorded and addressed, if possible. The Metals Advisory Group mailing list was expanded over time and is being used by RWQCP to communicate new program features.

In addition to the Metals Advisory Group, RWQCP held a series of workshops during 1990 and 1991 while the Silver Reduction Pilot Program was being developed and implemented. All potentially regulated facilities (approximately 500) were invited to the workshops. The attendance varied from 50 to 100 people. The following four types of workshop were held:

- Ordinance development workshops to allow for input before ordinance development;

- Ordinance description workshops to explain the adopted ordinance;
- Vendor workshops to allow vendors to provide information about their services and products that could help facilities comply with the ordinance; and
- Regulatory mechanics workshops to assist facility representatives with the administrative aspects of compliance with the ordinance, including permit applications, hauler certifications, and annual reports.

As documents were drafted that could be used by facilities (such as the mail survey of photoprocessing activities), the documents were tested on a group of volunteers. These volunteers were typically members of the Metals Advisory Group. This step saved time and frustration for all parties in the long run.

A newsletter titled *The Silver Lining* was written and sent to the affected facilities, the Metals Advisory Group, and other interested parties approximately every 6 months. It was used to announce new program components, answer questions from workshops that had to be researched, and issue friendly reminders about upcoming compliance deadlines. *The Silver Lining* was augmented by direct mailings between issues and media coverage and articles in other newsletters and publications.

Local Sewer Use Ordinance Requirements. Identical revisions were made to the local Sewer Use Ordinance in late 1990 and 1991 for the five member agencies within the RWQCP service area. The ordinance revisions established three principal objectives:

- A waste-minimization study authority,
- New local limit for silver, and
- Requirements for photoprocessors.

The ordinance was amended to provide authority for the member agencies to require waste-minimization studies from industrial and commercial facilities. While Senate Bill 14 provided the state with that authority and the existing ordinances were sufficient, new language was included in the ordinance making that authority clear to avoid debate and confusion.

A new local limit for silver was calculated based on meeting a maximum RWQCP effluent limit of 2.3 µg/L. At the time, that effluent limit had not been issued yet because of the NPDES permit directive by the Regional Board to conduct a pilot program. Nevertheless, the decision was made to base the local limit on 2.3 µg/L because RWQCP wanted to avoid having to return to the facilities for further reductions later. Subsequent to adoption of the ordinance, the Regional Board imposed the 2.3-µg/L limit.

In the first draft of the ordinance, the proposed local limit was 0.25 mg/L. This proposed limit was to apply to both industrial (mostly electronics and plating facilities) and commercial photoprocessing and X-ray facilities. Standard U.S. EPA techniques were used to calculate the allowable industrial and commercial flows. The flow rate used was the sum of total silver-bearing process flows from industrial sources and the total fixer (silver-bearing waste) flows from photoprocessing and X-ray development.

During the ordinance review process, photoprocessing industry representatives commented that the proposed 0.25-mg/L limit would be extremely difficult to achieve consistently for the mid- to small-volume photoprocessors. During the same time period, better information was obtained on process flows, in part from the mail survey of photoprocessors. In response to the new data and the comments, it was determined that the limit for photoprocessing could be raised to 0.5 mg/L as long as the limit for industrial processes would be held to

0.25 mg/L. This difference appeared reasonable in light of the much higher influent concentrations and the smaller facilities typical of photoprocessors and X-ray developers. These revised limits were ultimately adopted by the five member agencies.

REQUIREMENTS FOR PHOTOPROCESSORS. In addition to the 0.5 mg/L maximum limit, the ordinance specified a number of important program details:

- A choice was provided: either treat the silver-rich solutions (the solutions containing greater than 90% of the silver) or have them hauled off site.
- To promote water conservation and reduce the mass of silver, a limit of 1.0 mg/L silver applies to facilities with washless systems or achieving a 60% water reduction as compared to the manufacturer's recommendations.
- Washwater, estimated to contain much less than 10% of the silver, must be treated if the treatment option is chosen or if the facility generates more than 0.38 m^3 (100 gal) of spent fixer per month. This gives relatively small facilities the option of hauling off the fixer and not needing to attempt to treat the washwater. This compromise was deemed acceptable because of the small amounts of silver contributed by washwater from small facilities and the lack of cost-effective and practical treatment options.

Industrial Source Reduction Activities. The Silver Pilot Program for Industrial Facilities consisted of the following components: reduction in discharge limit, enhanced enforcement, waste-minimization studies, and metals-mass-discharge fees. Each of these components are discussed below.

REDUCTION IN DISCHARGE LIMIT. The most significant source control action taken by RWQCP may have been the reduction in the local limit from 5.0 to 0.25 mg/L. The methods of calculating this limit have been described above. However, the compliance rate for the metal-finishing industry (the primary group generating silver-bearing wastes) has historically been high. That silver-bearing waste streams are generally a small fraction of the total process flow from a metal finisher is part of the explanation.

ENHANCED ENFORCEMENT. In 1989, RWQCP began to implement an administrative penalties program designed to be a deterrent to violations. If violations are not corrected after a warning letter, action is escalated and a proposed compliance agreement is issued. Facilities are able to settle the matter by negotiating the signing of the agreement rather than becoming the subject of RWQCP litigation. The agreements typically contain monetary penalties roughly half of the maximum that could be levied by the court system. Fines peaked in 1990 (approximately $150 000) and are on the decline, as noted in 1991 (fines totaled approximately $75 000). It appears that a deterrent effect is occurring as fewer significant violations are occurring.

WASTE-MINIMIZATION STUDIES. The major industrial silver dischargers were identified on a mass basis and referred to as the Tier 1 industries for silver (five industries were found to contribute 85% of the industrial silver). This group was required to conduct waste-minimization studies first. If an enforcement action was underway, it was used as the vehicle for imposing the study requirement. If not, the permit reissuance process was used. Permit issuance schedules were adjusted so that imposition of all the study requirements occurred early in the program. Generally, 6 months was provided to complete the study. The ordinance, much like California's State Bill 14, does not require implementation of particular control measures but allows industry to decide which measures it will implement. The exception is when a facility is in violation; it must then take all reasonable steps for compliance.

The studies are reviewed, generally within 30 days, by RWQCP staff or its consultant. Typically, further information is requested and additional measures are suggested for further investigation. Site visits and one or more revisions of the study are common to encourage all reasonable pollution prevention measures to be taken.

It was envisioned that waste-minimization activities would be reviewed annually. Progress implemented would be reviewed and measures not implemented would be revisited.

METALS MASS DISCHARGE FEE. A fee schedule for metals has been adopted by RWQCP. The purpose is not to raise new revenue but to shift the current sewer fee schedule to reflect removal of metals as well as conventional parameters (biochemical oxygen demand, total suspended solids, and flow, for example). The metals fee schedule is intended to provide further incentive for reductions and for implementation of the waste-minimization studies. The initial fee was set at a low-level, $10 per pound (kilogram) of metals discharged. Once experience has been gained in administering the program, the metals fee would be raised and the traditional fees reduced accordingly.

Commercial Source Reduction Activities. With approximately 43 to 75% of the silver coming from commercial facilities, it was necessary to devote most of the program resources to this sector. No significant sources beyond photoprocessing (including X-ray development) were identified; thus, the commercial program focused on this activity. The activity was found to occur not only at photographic development shops, medical facilities, printers, and graphic artists, but also at a wide variety of schools, government buildings, businesses, and research facilities. Small photoprocessing operations were also found to exist at a number of currently regulated industries such as printed circuit board manufacturers. It was decided that the same rules would apply to all photoprocessors, whether the unit(s) existed at a small business or a large industrial complex already regulated by the industrial waste program. The newly adopted ordinance requirements would apply at the photoprocessing unit before dilution with other waste streams. As stated previously, photoprocessing facilities were given a choice: treating of the spent fixer or hauling it off site.

FACILITIES CHOOSING TREATMENT. Out of a total of 365, fewer than 20 facilities chose to treat, rather than haul off site, their silver-rich solutions. Silver-rich fixer is the principal, but not the only, high-strength solution of concern; the ordinance regulates "spent solutions" and includes other solutions that contain high silver concentrations. The facilities choosing treatment were issued a short permit—an abbreviated format from that issued to industrial dischargers. Simplicity was seen as the key. The compliance and the permit emphasized three key points:

- Discharge limit,
- Self-monitoring program, and
- Submission of an annual report.

For facilities generating more than 0.8 m^3 (20 gal) per month of silver-rich solution, a compliance period of approximately 9 months was established. The compliance period for smaller facilities was approximately 6 months. This concept allowed sufficient time from ordinance adoption to install a treatment system. Facilities generating less than 0.8 m^3 (20 gal) per month were encouraged to have their silver-rich solutions hauled off site for reclamation because treatment on site would likely not be cost effective for such small volumes.

FACILITIES CHOOSING HAULING OFF SITE. The vast majority of facilities, even those much larger than 0.8 m^3 (20 gal) per month, elected to have their waste hauled off site. In the early stages of the program, facility managers expressed concern about having wastes hauled away. Their concerns centered around whether they would become a responsible party for subsequent illegal disposal; accidental spills on the highway; and the availability of qualified haulers with the right insurance, permits, and approvals.

To address those concerns, a request for proposals was issued to solicit interest in being a hauler within the RWQCP service area. Two firms met the RWQCP requirements and were selected. These firms demonstrated that they were experienced; had the appropriate permits, approvals, and insurance; could offer a locally based, user-friendly program; and could publicize their services and keep records for the RWQCP in a cost-effective manner.

The hauling costs of both firms are similar. A photoprocessing business is charged a minimum of $150 for each pickup. Forty gallons are allotted from the hauler for storage and pickup. For volume more than 40 gal (0.15 m^3), a fee of $3.50 is charged per gallon (gallons $\times$ [3.785 $\times$ 10^{-3}] = m^3). Based on an average of 2 500 mg/L of silver in the spent solution, the cost is approximately $150/lb (lb $\times$ 0.453 6 = kg) of silver hauled.

Facilities choosing the hauling option submitted a hauling certificate to the RWQCP that documented their commitment to hauling rather than treatment. The RWQCP responded with a hauling acknowledgment letter. The ordinance requires facilities to maintain records of types and amounts of solutions generated and hauled and to submit an annual report on February 1 of each year.

SITE INSPECTIONS AND FOLLOW-UP. Following the September 31, 1991, compliance deadline, facility inspections commenced with the largest facilities inspected first. In the early months of the compliance phase, facility assistance, rather than enforcement, was the normal response to exceedances of the silver limit by small- and medium-size facilities. Enforcement was anticipated as the more typical response in future months.

REGULATORY STREAMLINING. For many photoprocessing facilities, the silver-reduction pilot program had the effect of exposing them to other regulatory agencies also focusing their attention to smaller commercial facilities. To assist these facilities in complying with all requirements simultaneously, RWQCP staff worked on regulatory issues with three agencies: California Department of Toxic Substances Control; Santa Clara County Hazardous Waste Management Program; and the local fire department.

CALIFORNIA DEPARTMENT OF TOXIC SUBSTANCES CONTROL. For medium- to small-size facilities, the California Department of Toxic Substances Control's permit-by-rule requirement was a deterrent to the installation of silver-treatment systems. Insurance and other administrative requirements were a roadblock despite the extremely small risk associated with the type of treatment system needed. The RWQCP staff actively supported legislation that resulted in an exemption for facilities producing less than 0.6 m^3/month (150 gal/month) of spent fixer.

SANTA CLARA COUNTY HAZARDOUS WASTE MANAGEMENT PROGRAM. Photoprocessing facilities discharging their spent fixer down the sink were "invisible" generators of hazardous waste. Shipping off site, with its attendant manifesting procedure, made them visible. Fees associated with the county's expanding Generator Permit Program were a deterrent to compliance. The RWQCP requested that the county ease the burden on small generators, and the county responded by reducing fees for the smallest generators.

LOCAL FIRE DEPARTMENTS. For facilities choosing to haul their waste off site, the fire department's hazardous material storage requirements were a roadblock to cost-effective compliance. If more than 0.04 m^3 (10 gal) of silver-rich solution was restored before off-site shipment, the permit requirements and fees were triggered. However, the economics of hauling were greatly improved if 0.1-m^3 (25-gal) containers were allowed for use. Because of the low health risk associated with storing silver-rich solution, the City of Palo Alto Fire Department raised its cutoff level to 0.11 m^3 (30 gal).

Residential Source Reduction Activities. Although the residential contribution was estimated to be relatively small, approximately 5% of the influent loadings, the need for residential reductions was included in the program. Some hobbyists were known to produce more silver-rich solution than certain businesses such as dental offices. It would be important to request reductions from residents if the program were to be perceived as equitable.

Environmental groups participating in the Metals Advisory Committee believed that the City of Palo Alto's existing Household Hazardous Waste Drop-off Program, conducted once a month, was not frequent enough to serve home darkroom hobbyists. Therefore, a variance was requested and obtained from the Department of Toxic Substances Control to increase the program from 1 day per month to 6 days per week for photochemicals. For this expanded program, the geographic area was also increased from including only the City of Palo Alto to including the entire RWQCP service area.

The expanded drop-off program was publicized in a variety of publications, including utility bill stuffers, newspapers, and newsletters. A point-of-sale brochure was developed and distributed to shops selling photochemicals. Shop owners and clerks were asked to place the brochure in the bag with photochemicals being sold. The brochure alerts the buyer to the silver problem in the San Francisco Bay and describes the residential drop-off program. All of the publicity for residents highlighted the problem of silver in the San Francisco Bay and the solution that was readily available to them.

Program Results

Initial results have indicated that the Silver Reduction Pilot Program has been extremely successful. Approximately 617 m^3 (163 000 gal) of silver-rich solutions are being regulated, 487 m^3 (128 600 gal) of which are being treated on site with the remaining 130 m^3 (34 700 gal) being hauled off site for silver reclamation. The decrease in maximum effluent concentrations from the last 4 years is compared with the limit of 2.3 µg/L. In the first 2 months of 1992, the maximum concentration measured (0.9 µg/L) was significantly below the effluent limits, down from a 1989 maximum of 14 µg/L. In 1990, the maximum silver effluent concentration was 9 µg/L, and in 1991, it was 2.4 µg/L.

Figure 14 illustrates the silver effluent concentrations from February 1989 through January 1992. During this period, influent wastewater was sampled via weekly, 24-hour composite samples. The effluent data is divided into three phases:

- Before regulation (through July 1990), phase 1;
- Implementation phase (August 1990 through September 1991), phase 2; and
- Compliance phase (October 1991 to February 1992), phase 3.

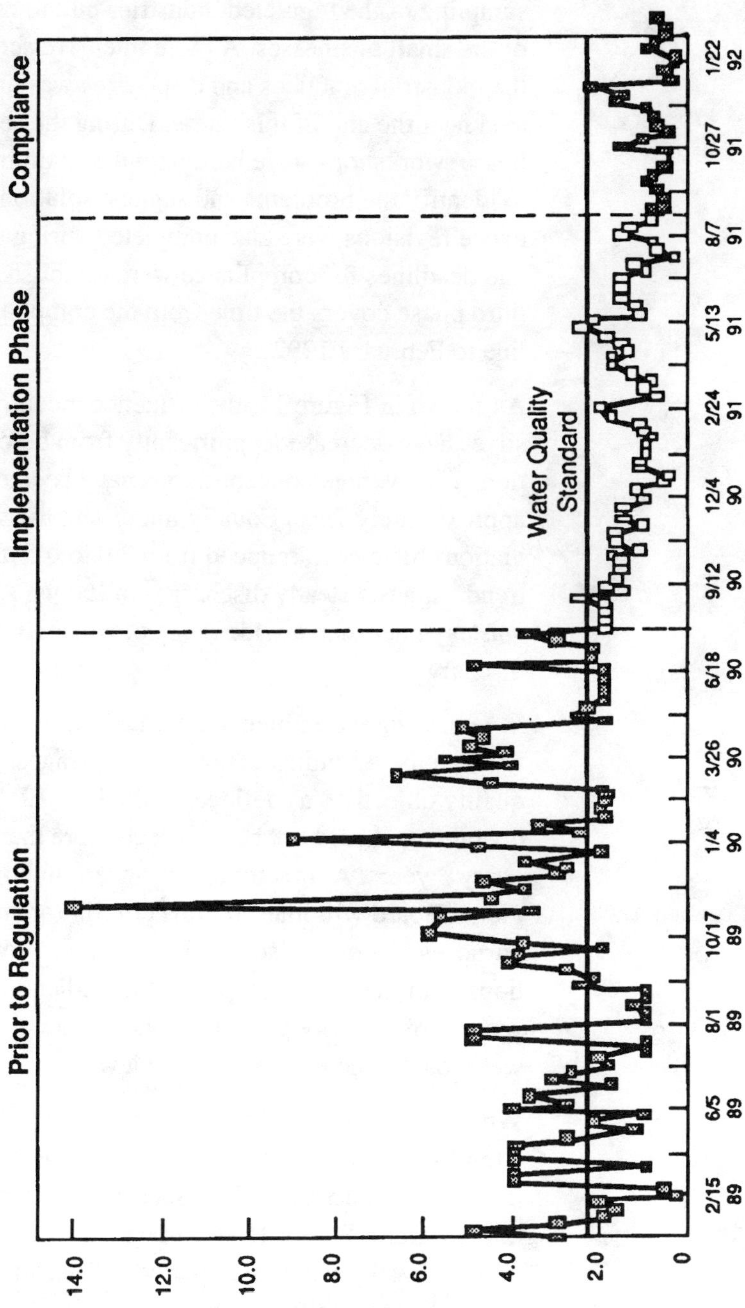

Figure 14 **Silver effluent concentrations, µg/L.**

The first phase was characterized by permitting and sampling of the regulated industries but no regulation of the small businesses. A more intensive scrutiny of the industrial practices and discharges was also initiated near the end of this phase. During the second phase, workshops were held with the small businesses to identify the problems and suggest solutions. Ordinance revisions were also completed during this time and deadlines for compliance were established. The third phase covers the time from the compliance deadline to February 1992.

As shown in Figure 14, the effluent concentrations of silver have decreased significantly from before regulation. The average concentrations have been reduced by approximately 75%. Equally important, the standard deviations have been reduced from 2.0 to 0.5 μg/L. This trend suggests steady discharges of known silver wastes rather than a wide range of concentrated batch discharges.

In reviewing the effluent data, maximum concentrations are better indicators than the averages. The water quality objective, as defined by the U.S. EPA criteria document, should not be exceeded more than once every 3 years. At this time, it is uncertain how the Regional Board will interpret this criteria for enforcement purposes. A comparison of the maximum concentrations with the limit may provide an indication of the success of the pilot program if the Regional Board were to take the most stringent view.

While the effluent results are encouraging, it must be emphasized that reduction in influent concentration is the true measurement of the success of a source reduction program. Figure 15 illustrates the influent concentrations for the same three periods. The same trends of reduced concentrations and lower variability can be easily seen. The averages and standard deviations of the silver influent concentrations for these three

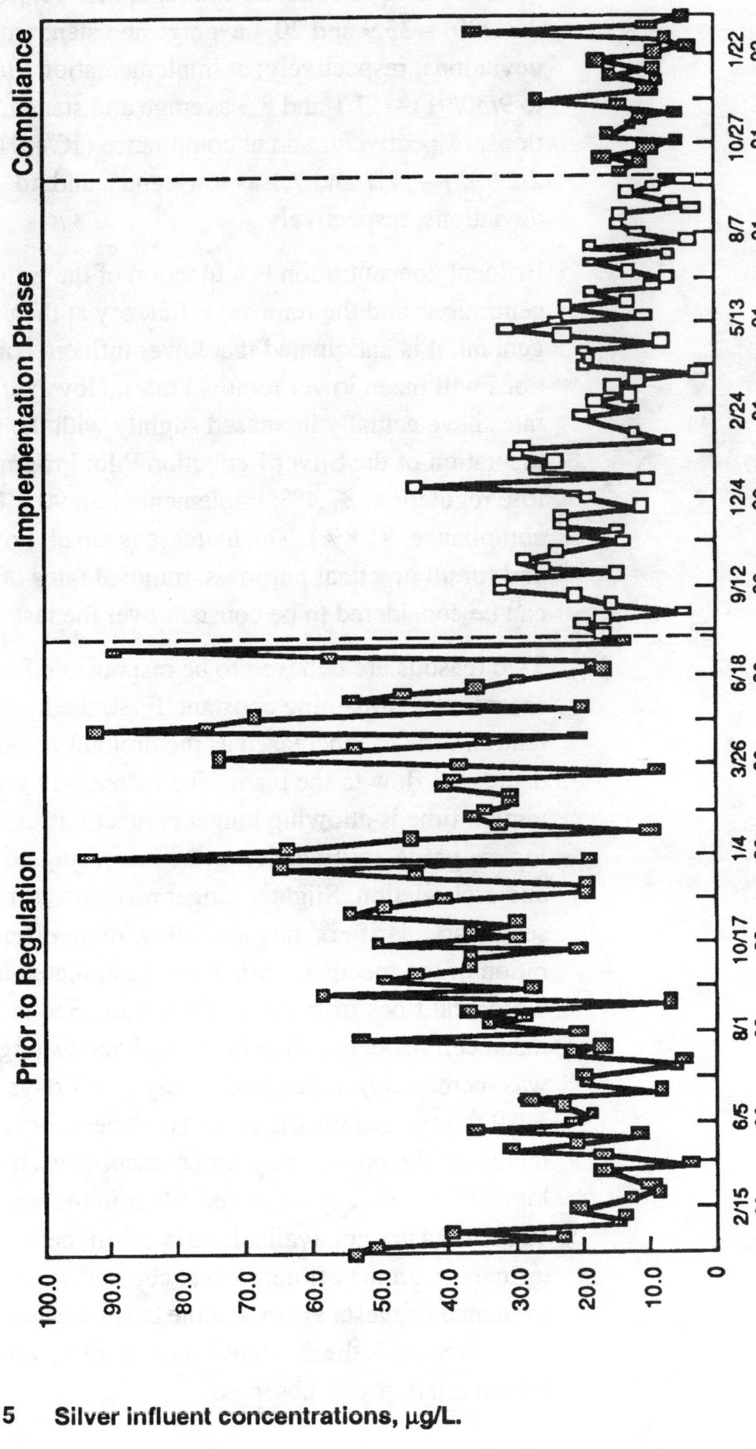

Figure 15 **Silver influent concentrations, µg/L.**

periods are as follows: before regulation (1/1/89 to 7/31/90)—33.9 and 20.4 average and standard deviations, respectively; at implementation (8/1/90 to 9/30/91)—17.1 and 8.3 average and standard deviations, respectively; and at compliance (10/1/91 to 2/29/92)—12.7 and 7.1 average and standard deviations, respectively.

Effluent concentration is a function of the influent concentrations and the removal efficiency at the plant. In general, it is anticipated that lower influent concentrations will mean lower removal rates. However, removal rates have actually increased slightly with the implementation of the Silver Reduction Pilot Program (before regulation, 87.4%; implementation, 90.3%; and compliance, 91.8%). The increase is small, however, and for all practical purposes, removal rates of silver can be considered to be constant over the last 3 years.

Two reasons are believed to be responsible for silver-removal rates remaining constant. First, the hydraulic retention time has increased as the drought response has decreased flow to the plant. The increased hydraulic retention time is allowing longer contact through the biological processes, which may be increasing adsorption and assimilation. Slightly longer residence time in the secondary clarifiers may also allow more efficient separation of the metals adsorbed and assimilated into the biological flocs from the liquid stream. Second, the mean cell residence time of the activated sludge system was increased from approximately 3 to 7 days in early 1989 to improve nitrification. This operational change increased the opportunity for contact between the biological floc and any dissolved silver in the wastewater. While data are not available to confirm these potential influences on the effluent silver concentration, their existence suggests again that the best measure of program success is the influent concentration, where a 63% reduction was observed.

One encouraging sign of success at RWQCB is a continual decline in silver concentrations in *Macoma balthica,* as shown in Figure 16.

Program Costs

Significant time and resources were required to implement the Silver Reduction Pilot Program. The total costs include expenditures from RWQCP as well as from the individual industries and commercial businesses. For the facilities choosing treatment, costs consist of capital as well as annual operating and maintenance (O & M) expenditures. For the facilities choosing hauling, the costs comprise hauling expenses. For RWQCP, effort during the first year was placed on identification of the contributors, solution development, and public information and participation. In the following years, the effort has gradually shifted into inspection and monitoring. The inspection efforts will be maintained in future years to ensure the continued success of this program.

The amortized capital and O & M costs for RWQCP and the industrial and commercial sectors are: industrial/commercial businesses section, $264 000; RWQCP, $146 000; and total costs, $410 000. Costs of the residential program are not shown separately as they are assumed entirely by RWQCP. In addition, some of the public education materials have been developed to educate both hobbyists and operators of commercial businesses and cannot be separated easily. The costs for RWQCP are further subdivided into the pertinent components and presented in Table 5.

Unit costs, in dollars per pound removed, for the two components of the Silver Reduction Pilot Program (RWQCP, industries, and commercial businesses) are calculated. The total unit costs of the program are compared to the installation and operation of a reverse osmosis unit at this plant. The unit costs of reverse

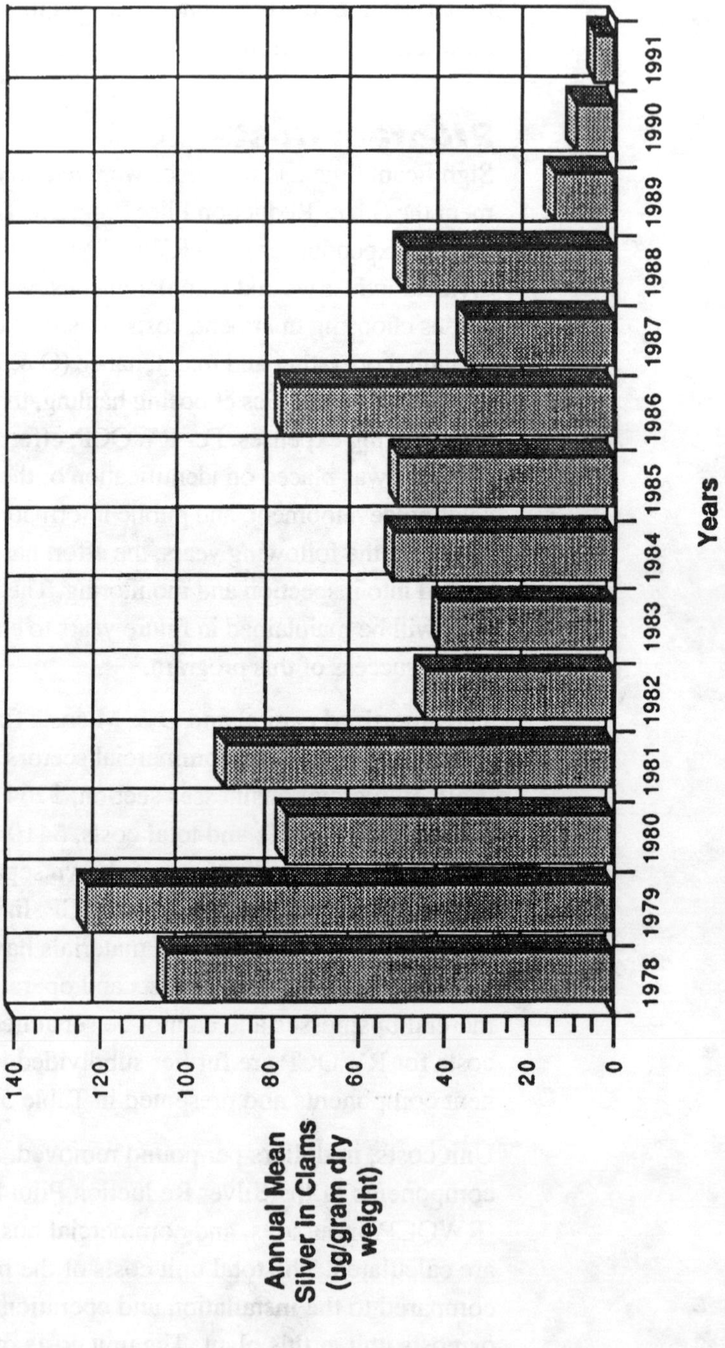

Figure 16 Silver concentration in *Macoma balthica* (clams) near Sand Point.

osmosis had been previously estimated to be approximately $2 700/lb of metals removed while the cost of the pilot program is approximately $320/lb of silver removed. The source control program for silver has been significantly more cost effective than end-of-pipe treatment at the RWQCP would have been.

Table 5 Silver-reduction program costs.

	One-time costs		Annual cost	
	Dollars	**RWQCP, person-years**	**Dollars**	**RWQCP, person-years**
Problem identification	$15 000	0.2	—	0.05
Ordinance development	$2 000	0.1	—	—
Public participation/information	$20 000	0.14	$6 000	0.1
Data handling	$10 000	0.2	—	0.3
Facility identification	—	0.3	—	0.1
Facility evaluations	—	—	—	0.2
Clerical	$5 000	0.2	$2 000	0.2
Permits/acknowledgments	—	—	—	0.1
Inspections/sampling	—	—	$40 000	0.1
Enforcement	—	—	—	0.1
Plant hauling program	—	0.05	—	0.01
Expanded residential drop-off	—	0.01	—	0.01
Totals	$52 000	1.2	$48 000	1.3

REFERENCE

Association of Metropolitan Sewerage Agencies (1993). Silver Discharge Survey. Final Rep., Washington, D.C.

ADDITIONAL SOURCES OF INFORMATION

Publications

Water Environment Federation (1993) Controlling discharges from small industrial/commercial sources. *Proc. Preconf. Sem.*, 66th Annu. Conf. Water Environ. Fed., Anaheim, Calif.

Agencies and Organizations

Silver Coalition
550 Mamaroneck Avenue
Harrison, NY 10528
(914) 698-7603

Office of Wastewater Enforcement and Compliance
U.S. Environmental Protection Agency
401 M Street, SW
Washington, DC 20460
(202) 260-7361

Institute for Research and Technical Assistance
2800 Olympic Boulevard, Suite 101
Santa Monica, CA 90404
(310) 453-0450

ACRONYMS

BMP	Best management practices
EDTA	Ethylenediaminetetraacetic acid
NPDES	National Pollutant Discharge Elimination System
O&M	Operations and maintenance
PDTA	Propylenediaminetetraacetic acid
SIU	Significant industrial user
SRU	Silver-recovery unit
TMT	Trimercapto-*s*-triazine
U.S. EPA	United States Environmental Protection Agency
WWTP	Wastewater treatment plant

INDEX

INDEX
(continued)